1986

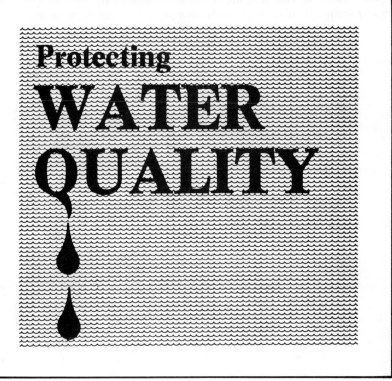

Protecting WATER QUALITY

Gary E. McCuen

IDEAS IN CONFLICT SERIES

publications inc.

411 Mallalieu Drive
Hudson, Wisconsin 54016

Illustration & photo credits
Carol and Simpson 46, 70, 171, Environmental Protection Agency 30, 54, 103, 116, 122, 128, 138, 167, Fish and Wildlife Service 23, 76, Government Accounting Office 96, Heller, West Bend News 172, Minneapolis Star and Tribune 11, 109, National Atmospheric Deposition Program 154, U.S. Geological Survey 16, 35, 162, 163, Water Pollution Control Federation 59, 83, 88, Wisconsin Department of Natural Resources 148

©1986 by Gary E. McCuen Publications, Inc.
411 Mallalieu Drive • Hudson, Wisconsin 54016 •
(715) 386-5662
International Standard Book Number 0-86596-056-9
Printed in the United States of America

CONTENTS

CHAPTER 3   GROUNDWATER CONTAMINATION

CHAPTER 4   PROTECTING SURFACE WATER

REASONING SKILL DEVELOPMENT

These activities may be used as individualized study guides for students in libraries and resource centers or as discussion catalysts in small group and classroom discussions.

IDEAS in CONFLICT ®

This series features ideas in conflict on political, social and moral issues. It presents counterpoints, debates, opinions, commentary and analysis for use in libraries and classrooms. Each title in the series uses one or more of the following basic elements:

Introductions that present an issue overview giving historic background and/or a description of the controversy.

Counterpoints and debates carefully chosen from publications, books, and position papers on the political right and left to help librarians and teachers respond to requests that treatment of public issues be fair and balanced.

Symposiums and forums that go beyond debates that can polarize and oversimplify. These present commentary from across the political spectrum that reflect how complex issues attract many shades of opinion.

A global emphasis with foreign perspectives and surveys on various moral questions and political issues that will help readers to place subject matter in a less culture-bound and ethno-centric frame of reference. In an ever shrinking and interdependent world, understanding and cooperation are essential. Many issues are global in nature and can be effectively dealt with only by common efforts and international understanding.

Reasoning skill study guides and discussion activities provide ready made tools for helping with critical reading and evaluation of content. The guides and activities deal with one or more of the following:

    RECOGNIZING AUTHOR'S POINT OF VIEW

    INTERPRETING EDITORIAL CARTOONS

    VALUES IN CONFLICT

    WHAT IS EDITORIAL BIAS?

WHAT IS SEX BIAS?

WHAT IS POLITICAL BIAS?

WHAT IS ETHNOCENTRIC BIAS?

WHAT IS RACE BIAS?

WHAT IS RELIGIOUS BIAS?

From across *the political spectrum* varied sources are presented for research projects and classroom discussions. Diverse opinions in the series come from magazines, newspapers, syndicated columnists, books, political speeches, foreign nations, and position papers by corporations and non-profit institutions.

About The Editor

Gary E. McCuen is an author and publisher of anthologies for public libraries and curriculum materials for schools. Over the past 14 years his publications of over 200 titles have specialized in social, moral and political conflict. They include books, pamphlets, cassettes, tabloids, filmstrips and simulation games, many of them designed from his curriculums during 11 years of teaching junior and senior high school social studies. At present he is the editor and publisher of the *Ideas in Conflict* series and the *Editorial Forum* series.

CHAPTER 1

THE AMERICAN WATER CRISIS: AN OVERVIEW

SURVEYING THE CRISIS

Robert A. Hamilton

Robert A. Hamilton is a newspaper reporter and a regular contributor to many national publications and magazines.

Points to Consider

1. Why may there be a shortage of drinking water in the near future?
2. What are the major threats to surface and groundwater?
3. Are any solutions suggested for water pollution?

Each year, 40 million tons of raw sewage, chemical wastes and other poisons pollute our water supply.

Before this decade is over, resource management experts say, the U.S. will realize how precious water really is. The "strategic shortage" of the '80s, they say, will be water—drinkable water.

"We're trying to alert people before the situation reaches crisis proportions," said Dr. Peter G. Bourne, president of Global Water, Washington, DC.

U.S. Geological Survey figures show that water use has more than doubled in the past 30 years, to about 450 billion gallons per day (BGD). That is about 2,000 gallons per person per day for every U.S. citizen. Even that demand, however, pales beside the amount of water available in this country: streamflow is about 1,200 BGD, and there are an estimated 33 quadrillion to 54 quadrillion gallons in underground water sources called aquifers that are replenished by rainfall and snow-melt with about 300 trillion gallons each year.

National Perspective

"From a national perspective, there is sufficient water to meet projected needs well beyond 1985," wrote Warren Viessman Jr., a senior engineer with the Congressional Research Office, and Christine Moncada, his assistant, in a report on state and national water-use trends to 2000.

"This optimism," Viessman wrote, "should be tempered, however, with the realization that national totals do not reflect geographic or temporal variations, and severe local, state and national problems can be expected."

The problem arises because about 75 percent of the U.S.' population lives on 2 percent of the land, and because much of the nation's food supply is raised on arid western lands that require extensive irrigation. Thus, there is some degree of inadequate surface water supply in all 21 of the country's water-resource regions, and moderate to extensive groundwater overdrafting (when water is withdrawn from an aquifer more quickly than it can be replenished) in 38 of 106 sub-regions.

Drinking Water

Yet, even if additional sources of water were somehow to be located, there is another and even more threatening problem: water pollution. Afflicting both surface and groundwater, pollu-

Reprinted with permission of *The Minneapolis Star and Tribune*

tion has severely limited the amount of drinking water available. Hazardous wastes have leached out of landfills; animal wastes, fertilizers, herbicides and pesticides have contaminated much agricultural land.

"We now recognize that a good deal of the precious drinking-water supply of this country in underground aquifers has been put in jeopardy by irresponsible disposal practices, and we are beginning to understand the enormous threat presented by the dumping of unprotected toxic wastes in unprotected landfills," said U.S. Rep. Albert A. Gore Jr., D.-Tenn., during a hearing before a House Science and Technology subcommittee.

In New Jersey, for instance, leakage of chromium, copper, zinc and other toxic metals and chemicals from industrial lagoons have contaminated millions of gallons of groundwater. New Jersey's Department of Environmental Protection, as part of a study conducted for the National Cancer Institute, surveyed public wells serving more than 1,000 customers. "The results of these tests are sobering," said Thomas Burke, director of the study. "The majority of the finished drinking waters contained low levels of potential cancer-causing volatile organics." These results, he continued, "clearly indicate the sensitivity of our water resources to the threat of chemical contamination. Evidence of toxic contamination was found in urban, industrialized areas as well as the most rural parts of the state."

Nor is New Jersey, although one of the states hardest hit by water pollution, alone in this respect. The U.S. Geological Survey estimates that six billion tons of hazardous wastes have already been dumped on U.S. land nationwide, and 40 million additional tons are being added every year. There are, the USGS says, about 7,000 hazardous-waste dump sites, another 200,000 chemical disposal sites, 200,000 municipal landfills and countless septic tanks, chemical spills and other threats to clean water across the country. Worse yet, more than 700 organic chemicals used in industrial processes have been detected in our underground water supplies, including 22 known cancer-causing agents, dozens of other toxic substances and many more that have not yet been tested for their effect on humans.

Industrial wastes, as bad as they are, are not the greatest threat to drinking water. That position is reserved for septic tanks, which are estimated to exude some 800 billion gallons of effluent into U.S. soils annually. The second greatest threat is posed by petroleum exploration and development, and the third greatest is landfills and dumps. The fourth-ranked threat is posed by agricultural chemicals, including dibromochloropane, a pesticide detected in 193 of 257 wells (including 20 public water-supply wells) tested in 24 California counties. Ethylene dibromide, a suspected carcinogen used as a fumigant for grains, citrus fruit and soils turned up in 20 percent of the wells tested in Florida.

Other contaminants include seawater and acid rain. On Cape Cod, for instance, some towns must post weekly warnings of the salt content in well water, because seawater is creeping into aquifers as the groundwater is slowly being exhausted. Acid rain has become so bad in some northeastern states that all fish and other animal life in the lakes have been killed, and planes dump lime on the ice during the winter to neutralize the water once melt occurs.

All in all, said the National Well Water Association and the USGS, about 1 percent of all U.S. groundwater sources—which provide drinking water to half the nation—has been contaminated. Even this small percentage is serious, said Philip Cohen, chief of the USGS water resources division, "because we generally find groundwater contamination in areas of the densest population and industrial activity. It is clear that the problem is going to get worse."

Water Management

Experts like Cohen agree that the first step in averting a water supply crisis is better water management—but management alone will not suffice. Ways of curbing water usage and of recycling water must be found.

One area in which tremendous savings might be realized, these experts said, is in controlling water waste. The first step is to repair or replace aging, leaky water supply systems in most industrial cities, some of which routinely pump 40 percent or even 50 percent more water than they can account for in billing. New York City, for instance, loses at least 100 million gallons of the 1.5 billion gallons it pumps every day through 6,000 miles of piping, half of which has been in place for 45 years or more. Three major water mains in the city have burst since August 1983, and the related costs—to say nothing of the wasted water—were staggering.

Industrial water use could also be cut by 5 percent through good housekeeping practices, said acting EPA administrator James N. Smith. Recycling could save even more, he said, and multiple use of water—such as the use in some parts of the west today of treated waste water for cooling systems run by major utilities. Up to 45 billion gallons of water per day could be saved by adopting innovative irrigation techniques, such as that developed in Israel which uses "drip irrigation" to produce almost twice the land's former yield of melons while using less water to do so.

"Water is the classic states' rights issue," Dr. Bourne said. But it's more than a legal argument now: it's a life-or-death issue for people, their cities, their animals and their industry. It's an issue that must be addressed quickly, before it's too late.

Our lives depend on it.

WATER IN AGRICULTURE: THE IRRIGATION DEBATE

Tristram Coffin

Tristram Coffin is the editor and publisher of The Washington Spectator, a national newsletter dealing with national and international political and social issues.

Points to Consider

1. What area of the world is the most heavily irrigated?
2. How much water is used in irrigation?
3. Why is water wasted in large irrigation systems?
4. What is the nature of California's water crisis?

Tristram Coffin, "The American Water Crisis," *The Washington Spectator,* March 15, 1985.

In 1950, the U.S. took 12 trillion gallons of water from the ground; by 1980 the figure more than doubled and is still going up. Each day, 21 billion more gallons of water flow out of water resources than flow in from rain, snow melt and water return.

The "fertile crescent" of the Middle East, now semi-desert, was heavily irrigated in Biblical times. During the rule of the Pharoahs and Ptolemies, water of the Nile was carried to large areas that now are little more than desert.

Today, the American Southwest is the most heavily irrigated area in the world, transforming a desert into a veritable garden of Eden.

The omen is: heavy irrigation may destroy the land by salt seepage and wipe out societies that grow up around the man-made oases. This is not an academic point. Three years ago, Senator William Armstrong (R-Colo.) warned: "The 1,400-mile Colorado River is the lifeblood of 17 million people, from Denver to San Diego. This river has made America's western desert bloom; in fact, 1.5 million acres of prime farmland are irrigated by it today.

"And yet, this magnificent river is being slowly poisoned as its waters become more and more saline; that is, adulterated by dissolved solids. Salinity is caused by two things: salt loading— which comes from contact with the very saline western soils and salty mineral springs—and by salt concentration, which is caused by evaporation and the increasing use of the river in the seven states it serves.

"At its headwaters, the Colorado River has less than 50 milligrams of salt in every liter of water; at Imperial Dam near the Mexican border, the number leaps to over 800 milligrams, an increase of more than 1,600%. At the turn of the century, this will reach a staggering 1,200 milligrams per liter. (The EPA's maximum safe-level for drinking water is 500 milligrams.)

"The salt load of 10 million tons annually which enters Lake Mead adversely affects more than 10 million people and one million acres of irrigated land."

The "Insidious" Problem

This is but one look at a staggering and "insidious" problem, one that could radically alter American life styles in a generation, drastically cutting food production, raising household costs and injuring health.

OFFSTREAM WATER USES

Thermoelectric Power
Industry
Irrigation
Public Supply
Rural

INSTREAM WATER USES

Hydroelectric Power
Fish and Wildlife
Recreation
Navigation

INCREASING DEMANDS FOR OUR FINITE WATER SUPPLY
Source: U.S. Geological Survey

Senator Dennis DeConcinci (D-Ariz.) explained to the National Press Club recently: "What's happening to water in America is more than an occasional accident, or even a series of isolated problems. The problem is more insidious than that. We are not running out of water, or even destroying it in the military sense. Water in America is steadily and too quickly becoming unusable.

"Water is becoming unusable because a lot of it is being contaminated, both above and below the ground. Water is also becoming unusable because delivery systems are old and falling apart, especially in the West, and because they can't be built fast enough to keep up with population shifts in the Sunbelt states."

The crisis is not just in the West. A study by the Army Corps of Engineers finds that population growth is dangerously increasing salt levels in the giant Chesapeake Bay. The "consumptive loss" of fresh water, by drawing fresh water from the tributaries, will rise from 500 million gallons a day to more than 2.5 billion by 2020.

Groundwater contamination has been found in every state and affects such cities as Little Rock, New Haven, Springfield, Ill., Pittsburgh, and Newark. Excessive leakage and water-main breakage has plagued major cities such as Boston, Houston and St.Louis.

The Solutions Ignored

Fortunately, there are solutions. They are drastic and expensive: curtail irrigation projects, enforce strict conservation, prohibit the use of toxic chemicals on farms and in factories, build modern water purification systems, and beef up research on removing salt from sea water. The cost and pressure from lobbies have simply passed on the problem to the next generation, when remedial measures may be too late. It is much easier and politically palatable to pour money into military adventure.

For example, President Lyndon Johnson had ready to go a $20 billion program to restore water and sewage plants. When he decided to escalate the Vietnam war, he abruptly cancelled the water program. The irony will not be lost on historians.

Today, the Reagan Administration, while asking $25 billion for

16

research on a Star Wars program that many scientists say won't work, has ignored the emerging water crisis. One is reminded of Shelley's lines:

"My name is Ozymandias, king of kings:
Look on my works, ye Mighty, and despair!"
Nothing beside remains. Round the decay
Of that colossal wreck, boundless and bare
The lone and level sands stretch far away.

Some Water Facts

A look at the facts is useful. Some 4.2 trillion gallons of water reach the U.S. in the form of rain or snow every year. About 92% of this evaporates immediately or runs off, unused, into the oceans. We withdraw some 400 billion gallons per day to irrigate, power and bathe America; 65% comes from freshwater sources such as lakes, rivers, marshes, reservoirs, springs; 20% from underground aquifers; and 15% from saltwater sources, such as inland seas.

About 83% of water consumed is used in farming, 8% in manufacturing, 7% in homes, and 1% each for power and on public lands. It takes 14,935 gallons of water to grow a bushel of wheat; 60,000 gallons to produce a ton of steel.

The sole source of drinking water for half of all Americans is underground water and, states former Interior Secretary Stewart Udall, "at least half of this is either contaminated or threatened with contamination," undermining health. The pollution is from

17

farm runoff of pesticides and herbicides, industrial chemicals, sewage and salt.

In 1950, the U.S. took 12 trillion gallons of water from the ground; by 1980 the figure more than doubled and is still going up. Each day, 21 billion more gallons of water flow out of water resources than flow in from rain, snow melt and water return.

(A study by the *Detroit Free Press* found that "waste and artificially low prices for water are the real problem in Arizona. It's not just the swimming pools, man-made lakes, unmetered sprinklers for lawns, particularly in Phoenix, and enormous fountains, including the biggest one in the world, which shoots water upward at 7,000 gallons a minute at the Fountain Hills desert development.

"It is also, in fact, agriculture, which uses 90% of consumed water. About three-fourths of it is in Central Arizona for crops whose production the government is limiting because of overproduction. Many Arizona farmers use the water on inferior land to grow such crops as sorghum and alfalfa that require large amounts of water. The rest of us are paying for it.")

The Irrigation Debate

As the rivalry between town and country for water grows more intense, a debate over widespread irrigation has come into the open.

Irrigation does expand acreage and produce bigger yields. However, Senator Dave Durenberger (R-Minn.) comments, "It makes no sense to spend billions of Federal dollars to irrigate semi-arid lands and then spend billions more to buy the crops because there are no markets." The *Washington Monthly* reports that "an array of tax breaks and farm subsidies" underwrites plowing and irrigating lands "ill-suited for crops. . . . Federal policy encourages enormous waste by providing water for irrigation at prices that cover as little as 2% of costs."

A General Accounting Office study finds that more than 50% of the irrigation water is wasted. Water is evaporated out of irrigation canals at a rate sometimes as high as 50%. Many large-scale irrigators use the huge center-pivot rigs that spray water into the air, instead of drip irrigation developed by Israel to save water.

"A hugely disproportionate share of the Federal irrigation program's benefits go to corporations running farms as large as 20,000 acres." (*Washington Post*) For example, in the San Joaquin Valley of California, water is provided for land owned by Getty Oil, Tenneco West and J. G. Boswell, a huge cotton corporation.

The water from Federal projects costs so little, comments the

18

Washington Monthly, that farmers find it cheaper to use than to save water. In California's Wetlands reclamation district—where the average farm is 2,400 acres and produces profits of half a million dollars a year—the Federal government is charging $10 per acre-foot. In neighboring areas, water on the free market may cost 100 times that amount. In South Dakota, users pay $3.10 an acre-foot for water that actually costs $131.50 to produce.

The Ogallala Aquifer

The real tragedy of the irrigation splurge is that it is using up water that is vitally needed for future generations. The depletion of the Ogallala aquifer is a case in point. Three years ago, *Time* reported, "The Ogallala aquifer, the vast underground reservoir of water that transformed much of the Great Plains into one of the richest agricultural areas of the world, is being sucked dry."

The aquifer stretches from South Dakota through Nebraska, where two-thirds of its water lies, to Wyoming, Colorado, Kansas, Oklahoma, New Mexico and Texas. "For the past three decades, farmers have pumped water out of the Ogallala as if it were inexhaustible. Nowadays, they disperse it prodigally through huge center-pivot irrigation sprinklers, which moisten circular swaths a quarter-mile in diameter. The annual overdraft—the amount of water not replenished—is nearly equal to the flow of the Colorado River."

A report by a Boston engineering firm, Camp, Dresser & McKee, estimates that by the year 2020 some 5.1 million acres of irrigated land will dry up. Some believe the report is too optimistic.

The effect on the national economy could be severe. Nearly 12% of our cotton, corn, grain, sorghum and wheat is watered by the Ogallala. Almost half the nation's beef cattle are fattened on high plains feedlots. In Texas alone, 70,000 water wells have been dug into the aquifer. Parts of the Panhandle have already used up more than half the water in the portion of the aquifer beneath them.

Farm manager Jim Bell admits, "We know we're losing our water. We've just got to learn to use it less—and better."

California's Water Crisis

Farther to the west, southern California is in the middle of a water crisis. The still-growing megapolises of Los Angeles and San Diego and the rich San Joaquin Valley that grows everything from oranges to cotton must import water from distances of hundreds of miles.

But that water will be reduced this year because of a Supreme Court decision, turning more water from the Colorado River to Arizona. At the same time:

• A new state population estimate says that by 2010 the California population will jump 10 million, from 23.8 million in 1980 to 34.4 million. This increase will mean an additional 3.5 million acre-feet of water needed yearly, mostly for urban areas.

• "In the San Joaquin Valley, groundwater overdrafts in excess of 2 million acre-feet a year have reduced underground aquifers in some areas to critical levels. Perhaps a million acres of farmland could go out of production without new surface water." (*Los Angeles Times*)

• *The Economist* reports: "The impure waste waters discharged into marshes and rivers from the western slopes of California's fruitful Central Valley" means that farming "may have to be abandoned." Water runoff with pesticides and such natural chemicals as selenium are accumulating in "unprecedented proportions."

Some Answers

Worldwatch Institute says that it is essential to raise "irrigation efficiencies" by 10%. This can be done by using "drip or trickle irrigation systems that supply water and fertilizer directly onto or below the soil." Experiments with drip irrigation in the Negev Desert show per hectare yields of increases up to 80% over wasteful sprinkler systems.

In desert areas of the Southwest, the heavy irrigation may give way gradually to greenhouse farming, which uses less water and produces higher yields. New growth forests and orchards could protect soil from drying out and act as rain forests.

The use of toxic chemicals in both farming and industry will have to be severely curtailed if America's drinking water is to be protected.

This is a greater threat to American well-being and prosperity than any external force. It deserves priority action by the Administration and Congress.

OUR ENDANGERED WETLANDS

U.S. Department of the Interior

The following comments are excerpted from a publication about America's wetlands by the U.S. Fish and Wildlife Service of the Department of the Interior.

Points to Consider

1. How many acres of wetlands have been destroyed between the 1950's and 1970's?
2. What benefits do wetlands provide?
3. What is the current state of America's wetlands?
4. What different kinds of wetlands are identified and described?

Our Endangered Wetlands, U.S. Fish and Wildlife Service, Department of the Interior, 1984.

The figures are staggering. They mean that an estimated 54 percent of the wetlands that existed in colonial times have vanished forever.

The term "wetlands" encompasses a variety of wet environments—coastal and inland marshes, wet meadows, mudflats, ponds, bogs, bottomland hardwood forests, wooded swamps, and fens. Throughout much of our history, such wetlands were regarded as foreboding, dangerous places which had little economic value. Indeed, over most of the past two centuries Americans have repeatedly enacted laws and devised programs that were aimed at encouraging the development of these areas. As a result, more than 100 million acres of the nation's wetlands have been destroyed. During the 20 years from the mid-1950's to the mid-1970's, such losses averaged 458,000 acres a year. The figures are staggering. They mean that an estimated 54 percent of the wetlands that existed in colonial times have vanished forever.

More recently, however, we have come to realize that wetlands are precious ecological resources—resources that nurture wildlife, purify polluted waters, check the destructive power of floods and storms, and provide all sorts of recreational activities. This new attitude is reflected by two decades of Federal and State laws and other programs that serve to preserve and protect our remaining wetlands.

The National Wetlands Trends Analysis

How much of the nation's wetlands remain today? What kinds of wetlands are most threatened? And which are disappearing at the fastest rate? These are the kinds of questions facing the Federal and State officials responsible for protecting precious wetland areas.

To answer such questions, the U.S. Fish and Wildlife Service (the primary agency charged with preserving our wetlands) launched a pioneering study called the National Wetlands Trends Analysis. For the first time, precise statistics show not only the nationwide status of wetlands but also trends in terms of gains and losses. All of this new information is needed to help officials develop or, if necessary, alter Federal programs and policies.

The massive study, which began in 1979, focused on the period between mid-1950's and the mid-1970's. Aerial photographs of wetlands taken in the 1950's were compared with pictures of the same areas taken some 20 years later. By measuring the

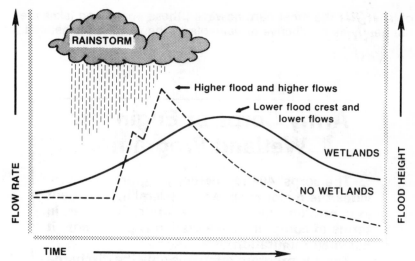

RAINSTORM

← Higher flood and higher flows

↙ Lower flood crest and lower flows

WETLANDS

NO WETLANDS

FLOW RATE

FLOOD HEIGHT

TIME

Wetland value in reducing flood crests and flow rates after rainstorms.
Source: Fish and Wildlife Service

changes that had taken place, analysts were able to answer a whole range of key questions with unprecedented statistical accuracy.

The State of America's Wetlands

In the mid-1970's, there was a total of 99 million acres of wetlands left in the United States (the lower 48 States). This represents about 5 percent of the nation's land surface. The overwhelming portion of these existing wetlands—93.7 million acres—consists of inland freshwater marshes, swamps, bogs, and ponds. The remaining 5.2 million acres are made up of coastal saltwater marshes.

Over the 20 years between the mid-1950's and the mid-1970's, net annual wetland losses averaged 458,000 acres. The hardest hit areas were inland marshes and swamps, the vegetated wetlands considered most valuable. During the two decades covered by the study, 6 million acres of forested wetlands, 400,000 acres of shrub swamps, 4.7 million acres of inland marshes and 400,000 acres of coastal marshes and mangrove swamps were destroyed. More than 11 million acres had disappeared—a total area twice the size of New Jersey.

The rapidly expanding demands of agriculture accounted for most (87 percent) of this great loss. The study indicated that 11.7 million acres of wetlands had been drained for crop production. At the same time agricultural development and construction projects created 2.1 million acres of ponds and 1.4 million acres

23

of lakes. For the most part, however, these ponds and lakes are not nearly as productive or valuable as the vanished vegetated wetlands.

Army Corps of Engineers Wetland Program

The corps' 404 regulatory program now provides the major avenue for Federal involvement in controlling the use of wetlands; however, in terms of comprehensive wetlands protection, it has major limitations.

The 404 program regulates only the discharge of dredged or fill material. Because projects involving excavation, drainage, clearing and flooding of wetlands are not explicitly covered by 404, these activities are not regulated by most corps districts. Yet, such projects were responsible for the vast majority of the wetland conversions that occurred between the mid-1950's and the mid-1970's. . . .

In the eyes of the corps, their primary function in carrying out the law is to protect the quality of water; protection of habitat and other wetland valves is only a secondary concern. Projects have been permitted under the 404 program regardless of their impact on wetlands because the Federal Government could not demonstrate adverse impacts to the quality of the nation's waters. In contrast to the corps' interpretation, Federal resource agencies and environmental groups feel that the mandate of the Clean Water Act obliges the corps to protect the integrity of wetlands, including their habitat values, and not just the quality of water.

In the absence of more direct Federal involvement, the conversion of most inland wetlands is likely to remain essentially uncontrolled.

Office of Technology Report, 1983

One important aspect not measured by the study was the deterioration of many wetlands. The reduced quality of wetlands stems from many causes, including pollution from rivers, streams and adjacent fields; urban encroachment; the building of highways and railroad roadbeds; the construction of ditches for mosquito control; and oil and gas development canals that allowed saltwater intrusion into freshwater marshes.

Major Findings of the Study

The Fish and Wildlife Service study involved careful measurements of a sample of the nation's wetlands. Therefore, most of the results identify national or broad trends (some also apply to migratory waterfowl flyways and specific States).

Significant Trends in Inland Wetlands and Lakes

—A net loss of vegetated wetlands (inland marshes, forested wetlands and shrub swamps) of 11 million acres, nearly all due to agriculture.

—An overall net gain in non-vegetated wetlands (ponds and inland mudflats) of 2.3 million acres, due largely to the building of farm ponds.

—A net gain in deepwater habitats (inland lakes and reservoirs) of 1.4 million acres resulting from construction projects.

Significant Trends in Coastal Wetlands

—A net loss in vegetated wetlands (coastal marshes) of 372,000 acres, mostly due to conversion into the open water of bays and sounds.

—An overall net gain in sub-tidal deepwater habitats (bays and sounds) of 200,000 acres.

Our Most Threatened Wetlands

Inland Marshes. The many thousands of shallow, waterfilled depressions that dot the upper midwest States and central Canada are called "prairie potholes." These gouge marks, left by the retreating glaciers of the last ice age, serve as a vast waterfowl breeding ground. The prairie pothole regions of the Dakotas, northern Montana, Nebraska, Iowa, Minnesota and southern Canada are a principal nesting area for North America's migratory waterfowl.

The prolific wetlands, however, are scattered throughout some of the nation's most fertile agricultural areas. Consequently, great amounts have been drained and turned into cropland. Nebraska's Rainwater Basin, for example, is a focal point in the central flyway that is used by millions of ducks, geese and cranes during their annual migrations. But 90 percent of the original wetlands in this basin have been destroyed.

The inland marshes of Florida provide both feeding areas for wading birds and wintering grounds for waterfowl. In addition, they supply breeding habitat for such species as rails, the mottled duck and the endangered Everglade kite. These wetlands are also prime habitat for furbearers, alligators, and various other kinds of wildlife. The conversion of the marshes to agriculture is significantly affecting both waterfowl and other wildlife populations.

Forested Wetlands. Nearly 80 percent of the 25 million acres of periodically flooded bottomland hardwood forests that once existed in the lower Mississippi Valley has been lost to agriculture. The remaining area still serves as the major over-wintering grounds for most of the continent's mallards and for virtually all of the wood ducks in the central United States. They also provide rich habitat for a wide range of other wildlife and spawning and nursery areas for fish. By 1977 only 5.2 million acres of bottomland hardwood forests were left in the Mississippi delta. Even today these remnants are shrinking as tracts are leveled, drained, and converted for (mainly soybean) farming.

The Pocosin wetlands of coastal North Carolina are covered with evergreen trees and shrubs. These wetlands serve a critical function by regulating the flow of fresh water to nearby coastal estuaries. This flow of fresh water is essential in maintaining the Pamlico Sound's productive fisheries. But the Pocosins are under great pressure from peat mining (for fuel) and, like other wetlands, for agriculture.

Coastal Wetlands. Coastal salt marshes and mangrove swamps are important wintering areas for waterfowl and breeding grounds for wading birds. They are also key spawning and nursery waters for most of our commercial and sport fish, including shellfish. Prior to the 1960's, such valuable coastal marshes (particularly in Louisiana, Texas, and Florida) were disappearing at a troubling rate.

The causes of the destruction have been both natural and man-made. Most of Louisiana's losses, for example, are due to sinking terrain and subsequent flooding. These conditions stem from a variety of factors such as a rise in the sea level, subsidence of the coastal plain, levee construction, channelization, and oil and gas extraction. Coastal losses that photo interpretation showed could directly be attributed to man have been due to urbanization—mainly dredging and filling for developments. Up until the mid-1970's, Florida coastal wetlands especially suffered from the effects of such development.

Future Prospects for America's Wetlands

Our precious wetlands can be preserved. The nation's coastal marshes, for example, are faring better than other types of wetlands because of protective laws enacted by the Federal government and a number of States during the 1960's and 1970's.

However, as the space needed for cropland continues to grow and urban areas continue to expand, America's wetlands will inevitably shrink. Therefore, it is more important than ever to monitor wetlands and proposed alterations of wetlands. Such vigilance is required to provide information that officials need for making wise decisions.

The U.S. Fish and Wildlife Service is working, through a variety of programs, to conserve our existing wetlands. These programs include Federal permit reviews, waterfowl habitat acquisition, wetlands preservation easements, and environmental education. A 1982 national survey by Louis Harris revealed that a majority of the public supports such programs. Harris' *Survey for National Resources Council of America* found that 83 percent of those surveyed feel it is "very important" to preserve the nation's remaining wetlands.

HAZARDOUS WASTE SITES AND SUPERFUND

The Government Accounting Office

The following comments were reprinted from a report on hazardous waste sites and the progress of the Superfund clean up program. The report was issued by the Comptroller General of the Government Accounting Office.

Points to Consider

1. How is the Superfund Act and Program described?
2. How successful has it been in cleaning up waste?
3. What dangers do major hazardous waste sites pose?
4. How many hazardous waste sites have been located and what will it cost to clean them up?

The Comptroller General, "Cleaning Up Hazardous Wastes: An Overview of Superfund Issues," *Report to the Congress of the United States,* Government Printing Office, March 19, 1985.

Although EPA estimates that its inventory of potential sites may grow to 25,000, it could dramatically increase the program's size to over 378,000 additional sites.

Toxic chemicals at thousands of hazardous waste sites across the country continue to seep into the nation's groundwater, contaminate the land, and poison the air. The 1980 Superfund Act sought to control this threat by providing a $1.6 billion cleanup fund accumulated largely from taxes on petroleum and certain chemicals. The taxing authority for this program expires in September 1985; this provides the Congress with the opportunity to assess the program's status and direction. . . .

Background

The Superfund program represents a departure from most other environmental laws. Laws such as the Clean Water and Clean Air Acts give the Environmental Protection Agency (EPA) responsibility for setting national standards and ensuring compliance. States are often delegated authority to administer these environmental programs, with EPA exercising oversight.

Under Superfund EPA cleans up the nation's worst hazardous waste sites (called priority sites) and, to the extent possible, requires the parties responsible for these hazardous conditions to either perform cleanups themselves or reimburse the government for cleaning up the sites. Although EPA has responded to short-term emergency situations at non-priority sites, it considers cleanups at such sites to be primarily a state or local responsibility.

Although uncontrolled hazardous waste sites pose a substantial danger to human health and the environment, the scope of the hazardous waste problem, the degree of health risks involved, and the cost of correcting these problems are unknown. As of December 31, 1984, EPA had identified 19,368 hazardous waste sites, of which 538 have been designated as priority sites. Of these, 194 have no cleanup action currently underway or in the planning stage.

Under Superfund EPA has no mandate to set nationwide cleanup standards or oversee state-conducted cleanups. The absence of standards complicates an already lengthy, complex process for cleaning up hazardous waste sites. During Superfund's reauthorization, federal and state roles and responsibilities may need to be reassessed.

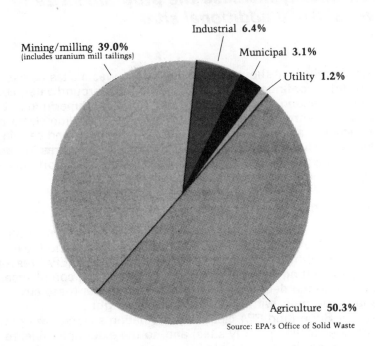

**6 BILLION TONS OF SOLID AND HAZARDOUS WASTE
ARE GENERATED IN THE U.S. EACH YEAR**

(Excludes high-level radioactive waste)

Industrial **6.4%**

Mining/milling **39.0%**
(includes uranium mill tailings)

Municipal **3.1%**

Utility **1.2%**

Agriculture **50.3%**

Source: EPA's Office of Solid Waste

Principal Findings

EPA and the states have given site discovery relatively little emphasis. Although EPA estimates that its inventory of potential sites may grow to 25,000, it acknowledges that a systematic discovery effort and a change in program emphasis toward cleaning up sites that received less emphasis earlier in the program, such as mining-related sites, could dramatically increase the program's size to over 378,000 additional sites.

The precise nature of the health risks posed by hazardous waste sites is also unknown. The Department of Health and Human Services has completed few Superfund-mandated health studies on the relationship between toxic substances and illness.

EPA estimates that federal cleanup costs for priority sites in 1983 dollars could range from $7.6 billion to $22.7 billion and that these cleanups could take until fiscal year 1999. This estimate depends on EPA's basic assumptions concerning the number of sites needing cleanup, the average cost per site, and the level of responsible party contributions to site cleanup.

30

1.3 Million Toxic Sites?

EPA estimated that as many as 1.3 million sites might have to be discovered and evaluated to determine if they are problem sites.

General Accounting Office, March 26, 1985.

Using alternative assumptions based on historical data and other available information, GAO projects that Superfund costs for priority sites in 1983 dollars could range from $6.3 billion to $39.1 billion and that these cleanups could take until fiscal year 2017. Also, the Department of Defense estimates that cleanup of its sites could cost an additional $10 billion. Since Superfund by law cannot be used to clean up federal facilities, Defense's costs must be funded through its budget.

"How Clean is Clean?"

The long-term cleanup process involves a series of activities such as performing technical studies and designing and implementing cleanup projects. EPA considers 10 sites to have received final cleanup action, and expects to take long-term cleanup actions at about 10 percent of the 19,368 known sites.

Neither the Superfund Act nor EPA regulations define hazardous waste site cleanup standards. Available solutions range from no action, to temporary containment of wastes, to total elimination. The option selected depends on the cost-effectiveness of the solution in relation to funds available for other site cleanup actions.

In the absence of specific hazardous waste standards, EPA considers applying other environmental laws in determining the extent of site cleanups. These other laws, however, do not address all of the substances and conditions found at sites. Part of the difficulty in setting standards is that little information is available on the risks posed by chemicals at these sites.

Federal Limitations

Except for emergency actions, EPA limits its cleanup efforts to priority hazardous waste sites. Although EPA has not taken an active role at non-priority sites, some state governments have programs to clean up these sites. State resources, authorities, and capabilities, however, vary. This variance, coupled with the absence of cleanup standards, may result in the public's not receiving uniform protection from the dangers of hazardous wastes.

During reauthorization deliberations several alternatives, including the following, are available for structuring the act:

(1) Make no change in the basic structure of the act. Superfund would continue to provide for cleanup at only the nation's worst hazardous waste sites on a priority basis, as resources will allow. EPA would not have responsibility for setting national standards or delegating cleanup functions to the states.

(2) Change the structure of Superfund more along the lines of previous environmental legislation, emphasizing permanent, long-term remedies and giving EPA responsibility for setting national standards for all hazardous waste sites. States could be delegated some or all cleanup functions, with EPA retaining oversight responsibility.

The information GAO has developed suggests that the Congress should consider the merits of changing the act's structure. The absence of national cleanup standards complicates an already lengthy, complex process for cleaning up hazardous waste sites. The lack of precise data on the health and environmental effects of hazardous waste sites makes standard setting difficult. Nevertheless, if we are to provide consistent site cleanup on a national basis, it is important that, where feasible, reasonably uniform criteria be established to govern both federal and state cleanup decisions.

CLEAN WATER LEGISLATION: HISTORY AND OVERVIEW

League of Women Voters

The following comments were excerpted from Blue Print for Clean Waters, *a pamphlet written by Deborah A. Sheiman, staff specialist, Natural Resources Department of the League of Women Voters Education Fund.*

Points to Consider

1. How does the Clean Water Act work?
2. What do the terms BAT, BCT, BPT and NPDES stand for and how are they defined?
3. How has progress been made under the Clean Water Act?

Every citizen has a personal stake in the quality of the nation's waters. We need water that is safe to drink, safe to swim in, habitable for aquatic life, free of nuisance conditions, and usable for agriculture and industry. Health, jobs, and the quality of our lives are thus affected in many ways by water quality.

Environmental Quality—1977
The Eighth Annual Report of the Council on Environmental Quality

Water pollution means lakes and streams choked with algae and weeds, wildlife habitat destroyed due to the drainage and filling of wetlands, beaches closed to swimming and wells poisoned by toxic chemicals. . . .

A federal program to address some of these problems in a comprehensive fashion was established in 1972 with the enactment of Public Law 92–500, the Federal Water Pollution Control Act Amendments. This law was the product of long experience with generally ineffectual state and federal pollution control programs. In 1977, following a study by the National Commission on Water Quality, the law underwent a series of "mid-course corrections," and its name was changed to the *Clean Water Act (CWA)*.

Though water quality problems persist, progress has been made—evidence that the Clean Water Act is working. It has prevented further deterioration of water quality during a decade of economic and population growth. It has helped some waterways recover to the point that fish are returning after a quarter century's absence and swimming beaches are reopening. But there is still a long way to go before long-term aquatic productivity is ensured. Improvement cannot continue unless the Congress, the Administration, officials at every level of government, industry and the public renew the commitment to clean water they made a decade ago. . . .

The Reagan administration's budget cuts and regulatory reform program inevitably mean reduced implementation and enforcement. And, at the same time that grants to states are being cut back, there is a move afoot to transfer more program responsibility from the federal level to the state governments. . . .

34

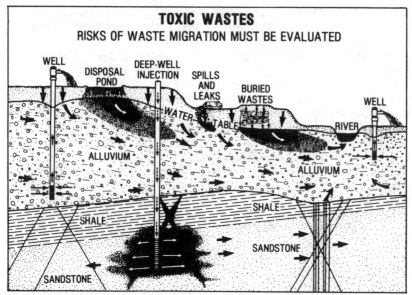

TOXIC WASTES
RISKS OF WASTE MIGRATION MUST BE EVALUATED

Source: U.S. Geological Survey

How the Clean Water Act Works

The regulatory approach of the 1972 act was based on unsatisfactory experience with earlier laws. Before 1972, pollution control efforts relied on a seemingly logical, but in practice faulty, state-by-state approach. Each state designated "beneficial uses" of water bodies (such as swimming, fishing or navigation), established numerical water quality criteria that set forth the conditions necessary to support those uses, then tailored specific water quality standards to individual streams or stream segments as the basis for enforcement actions.

Of the many problems inherent in this approach—including inadequate standards, funding and research—the worst was in enforcement. State officials had to negotiate with and among polluters along a waterway to achieve water quality standards. Industries could threaten to pack up and move to another state if they didn't like the compliance program. . . . The least workable aspect of the pre-1972 approach, however, was its underlying assumption that waterways have the capacity to assimilate wastes. The 1972 amendments were based on a new philosophy: that, as stated in the Senate Committee report, "no one has the right to pollute . . . and that pollution continues because of technological limits, not because of any inherent right to use the Nation's waterways for the purpose of disposing of waste."

The 1972 law states an ecologically based objective—"to re-

35

store and maintain the chemical, physical, and biological integrity of the Nation's waters"—and sets forth some specific goals for meeting it:

• that the discharge of pollutants into navigable waters (broadly defined as the waters of the United States) be eliminated by 1985 (the "zero-discharge" goal); and

• that, wherever attainable, an interim goal of water quality that provides for recreation in and on the water be achieved by July 1, 1983 (the fishable-swimmable goal).

The goals of the Clean Water Act are just that; they are not meant to be enforceable deadlines.

The Clean Water Act is designed to protect the Chesapeake Bay, the Great Lakes and the delicate ecology of all the rivers, lakes, streams, marshes, estuaries and oceans in the United States. To make progress toward meeting its ambitious objective, the 1972 law applied a mix of "carrots" and "sticks" to the major sources of pollution: *point sources* such as municipal and industrial outfalls, and *nonpoint sources* such as agricultural and urban runoff and dredge and fill operations. The most significant concepts are outlined below.

Water Quality Standards—The law retained the concept of water quality standards. EPA is required to establish *water quality criteria* that state the concentrations of various constituents necessary to support various uses. States use the criteria as guidance in setting specific *water quality standards* for all bodies of water. EPA regulations spell out an "antidegradation" policy that requires water quality standards to protect high-quality waters from degradation and requires states to review their water quality standards every three years and upgrade them with a view toward meeting the fishable-swimmable goal.

Control of Discharges (Effluent Limits)—The act required nationwide technology-based controls on the sources of pollution—specifically, new and existing "point sources" such as industries or municipal wastewater treatment plants that discharge through a pipe or culvert. EPA establishes *effluent guidelines* for major industries, a prescribed base-line level of pollution reduction required for each plant. Permit writers (see next section) must consider the effluent guidelines as well as the water quality standards in arriving at the *effluent limitation,* which is a specific control requirement placed on individual polluters. This limitation must be at least as strict as the effluent guideline, regardless of the quality of the receiving waters, but may be stricter if warranted by the state's water quality standards.

Enforcement by Permit—The 1972 law added an enforcement tool: a permit system. Point source polluters must get a permit under the *National Pollutant Discharge Elimination System (NPDES).* Permits specify the maximum allowable amounts and

Fish Kills

Because the reporting program is a voluntary one, only a fraction of the total kills are recorded. Even so, the economic losses traceable to these reported cases are staggering: from 1960 to 1979, over 580 million fish, including many with sport or commercial value, have been reported killed, largely by industrial, agricultural and municipal operations. In 1979, the number reported, 6.4 million, was the lowest since 1960. More significant than numbers of fish killed is a loss of a species or a decrease in species diversity, trends that are not systematically monitored.

League of Women Voters, 1983

constituents of effluent, a schedule for compliance and self-policed monitoring and reporting requirements. Authority to issue NPDES permits rests with EPA, which so far has delegated this responsibility to about 35 state governments. A separate permit system regulates the discharge of dredged or fill materials into wetlands and other waterways.

Ocean discharges were regulated in a slightly different manner. Under EPA guidelines, dischargers of pollutants that will cause unreasonable degradation of the marine environment are denied the necessary NPDES permit. . . .

Construction Grants Program

The 1972 Clean Water Act required secondary treatment of wastewater by July 1, 1977. Only one-third of the country's municipalities met the deadline. The 1977 amendments extended the deadline to July 1, 1983 (and waived the requirement for municipal dischargers into deep ocean waters, if certain environmental criteria could be satisfied).

To help local governments meet the nation's clean water goals and catch up with the backlog of municipal pollution problems, the Clean Water Act has authorized generous federal grants for building wastewater treatment facilities. EPA was empowered to pay 75–85 percent of the costs to expand existing facilities or to plan and build new *publicly owned treatment works (POTWs)*. State and local governments pay the balance of the construction tab. The 1977 amendments *authorized* $4.5 billion for fiscal year (FY) 1978, and $5 billion annually for 1979 through 1982. Actual appropriations were somewhat less. In 1981, for example, $3.5 billion was *appropriated,* but President Reagan limited actual *funding* to $2.4 billion.

This "section 201" program, as it is sometimes called, is one of the largest public works programs in American history, second only to the interstate highway system. Like the highway program, it has had far-reaching environmental impacts and has been subject to controversy. Some observers believe that large federal grants for excess capacity and collector sewers to feed central plants have induced urban and suburban sprawl. Environmentalists have criticized the subsidies for encouraging increased development of sensitive natural areas such as barrier islands.

Since 1972, over $25 billion in federal funds have been given out to build new sewage treatment plants or to expand or modify existing ones, but only a fraction of the projects have been completed. As of September 1980, only 2,223 (mostly small) treatment projects, valued at $2.8 billion, were on line. About 10,000 others are in the planning or construction state, with total time for completion of projects averaging eight years.

Equally disturbing is the poor performance record of the operating plants. EPA estimates that between 50 and 75 percent of them are in violation of their NPDES permits, and a General Accounting Office (GAO) investigation came up with even more alarming data. The principal performance problems stem from inadequate design and equipment, infiltration/inflow overloads, toxic industrial inputs that wipe out the populations of essential bacteria, and deficient operation and maintenance.

Communities pay to operate and maintain their treatment works, once they are built, through user charges determined by metering water consumption or through local property taxes. Large industrial users pay a fee based on the quality as well as the quantity of their wastewater.

Shortly after taking office in 1981, the President put a $2.4 billion spending cap on the construction grants program for FY 1981 (even though $3.5 billion had been appropriated) and included no funding in the 1982 budget.

With the program dead for lack of funding as of October 1, 1981, project construction came to a halt in those 30 or so states that had used up their grant monies. . . .

The Clean Water Act distinguishes among several kinds of industrial dischargers: direct sources that legally discharge effluent straight into water bodies; indirect sources that discharge into sewer systems leading to a municipal wastewater treatment plant; runoff from industrial sites; and accidental spills of oil and hazardous materials into water.

Direct Discharges

By 1977, 80 percent of the nation's 54,000 industrial sources had met the first benchmarks of the 1972 act—use of best *practicable* technology for pollution control—with stricter best *available* technology slated for 1983. But the system had two major flaws: some of the worst polluters were not in compliance, and only six toxic pollutants were regulated.

The Natural Resources Defense Council (NRDC) and other environmental groups sued EPA a number of times in an attempt to force better control of toxic water pollutants. The resulting judicial decision, known as the 1976 "NRDC consent decree," changed the approach of EPA's regulatory program for industrial dischargers. The slow chemical-by-chemical approach (which put an enormous burden of proof on EPA) was replaced by a technology-based scheme for regulating 65 classes of "priority" toxic pollutants from 21 major industries. In 1977, Congress incorporated this decision, with some extensions of deadlines, into the Clean Water Act.

EPA has refined the 65 classes of pollutants listed in the consent decree to include 129 specific chemicals and has updated the original list of 21 industrial categories to 34 major industries. The system is supposed to work on the basis of industry-specific, technology-based effluent guidelines, enforced by the NPDES permit system.

Different levels of control are required for different kinds of industrial discharges. Though confusing at first glance, the acronyms represent the nuts and bolts of the Clean Water Act:

● *best available technology economically achievable (BAT)* for toxic discharges from existing sources, with compliance required by July 1, 1984;

● *best conventional technology (BCT)* for discharges of conventional pollutants (see chart) from existing sources, with compliance required by July 1, 1984;

● a BAT pretreatment standard for industrial discharges into municipal sewer systems; and

● a new source performance standard (*best available demonstrated control technology*) for new facilities.

● For nonconventional pollutants, *best practicable technology (BPT)* was required by April 1, 1979 (or July 1, 1983, if the industry has contracted to discharge to a publically owned treatment works (POTW) that cannot yet accept the discharge). The more stringent BAT is required for nonconventional pollutants by July 1, 1984, or within three years of issuance of effluent guidelines, but no later than July 1, 1987. Variances for nonconventional pollutants are allowed: BPT is permitted, if dischargers can show that this is the maximum level economically feasible or that such a level is environmentally acceptable. . . .

The effectiveness of these industrial pollution control provisions depends to a large extent on how vigorously they are enforced. . . .

In July 1981, a federal court overturned EPA's rules for conventional pollutants on the grounds that they were too costly. If the ruling is not reversed in a higher court, EPA will have to start from square one and rewrite the BCT rules for industry.

Within a week of the court ruling on conventional pollutants, EPA filed a motion in the U.S. District Court seeking a relaxation of its obligations to control toxic industrial pollutants under the NRDC consent decree. Citing personnel and budget cutbacks, EPA sought to extend further the deadlines for issuance of effluent guidelines for industry, limit the number of industries covered, be released from its obligation to identify and regulate the many other toxic pollutants that were not included on the priority list of 65, delete the requirement that EPA publish water quality criteria for toxic pollutants (though this has already been done for 64 of the 65 classes of pollutants), and drop the development of a program to deal with toxic chemical "hot spots" where "better-than-BAT" controls are needed to protect against toxic effects. EPA has identified 34 streams bordered by major industrial complexes that are potential toxic hot spots.

Pretreatment and Indirect Discharges

The now infamous poisoning of the James River in Virginia and portions of the Chesapeake Bay with the highly toxic pesticide Kepone was due in part to an "indirect discharge." A manufacturing plant no bigger than a gas station was discharging Kepone into the sewer system of the small town of Hopewell, Virginia. The poison wiped out the sewage treatment process at the plant and was discharged into the James River along with all the untreated sewage, ruining the area's once prosperous fishing industry. This disaster demonstrates the importance of a pretreatment program for industries that discharge into public sewer systems.

In many cases, municipal sewage is helpful in breaking down organic industrial wastes, and joint treatment systems can offer both economies of scale and increased efficiency. But where toxic constituents are present, pretreatment is necessary to ensure that the Hopewell experience is not replicated. Without pretreatment, toxic substances can pass through the municipal facility untreated and be discharged in its effluent or contaminate the sludge, precluding its use for beneficial purposes.

GLOBAL WATER SUPPLIES: EXAMINING COUNTERPOINTS

This activity may be used as an individualized study guide for students in libraries and resource centers or as a discussion catalyst in small group and classroom discussions.

The Point

In the previous decade or so, water experts have concluded that the "likelihood of the world running out of water is zero." The recent UN Report of the World Environment, for example, tells us not to focus upon the ratio between physical water supply and use, and emphasizes making appropriate social and economic as well as technological choices. From this flows "cautious hope from improved methods of management." That is, an appropriate structure of property rights, institutions, and pricing systems, together with some modicum of wisdom in choosing among the technological options open to us, can provide water for our growing needs at reasonable cost indefinitely.

Julian Simon and Herman Kahn, *The Resourceful Earth,* 1984.

The Counterpoint

By the year 2000 population growth alone of the world will cause at least a doubling in the demand for water in nearly half the countries of the world. . . . Much of the increased pressure will occur in the developing countries when, if improved standards of living are to be realized, water requirements will expand several times. Unfortunately, it is precisely these countries that are least able, both financially and technically, to deal with the problem.

It is estimated that about half of all the irrigated lands of the world have been damaged by salinization, alkalinization and waterlogging. . . .

As pressures on water resources increase, conflicts among nations with shared water resources are likely to intensify.

Global 2000 Report To The President, 1980.

Guidelines

Social issues are usually complex, but often problems become oversimplified in political debates and discussion. Usually a polarized version of social conflict does not adequately represent the diversity of views that surround social conflicts.

1. Examine the counterpoints above. Then write down other possible interpretations of this issue than the two arguments stated in the counterpoints above.

2. Do you agree more with the point or the counterpoint? Why?

3. Which opinion in this chapter best illustrates the point?

4. Do you find any reading in this chapter in basic agreement with the point or the counterpoint?

5. Why can there be such basic disagreements over the issue of our global water supplies?

CHAPTER 2

BLUEPRINTS FOR CLEAN WATER: IDEAS IN CONFLICT

READINGS

43

A STRONG FEDERAL PROGRAM: THE BLUEPRINT FOR CLEAN WATER

Izaak Walton League

The following comments were made by Daniel Weiss, a conservation associate specializing in water pollution and water quality issues for the Izaak Walton League of America.

Points to Consider

1. Why should the federal government have a primary responsibility for setting and enforcing national water pollution controls?
2. What benefits have the nationwide programs under the Clean Water Act bestowed upon states?
3. What changes did the Environmental Protection Agency propose?

Excerpted from a public statement by the Izaak Walton League of America, February 23, 1983.

The nationwide program has enabled many states—including Minnesota, California, New Jersey, and Maryland—to develop and implement tough clean water programs without fear of losing industry to states with weaker programs.

The Clean Water Act, passed in 1972, gave the federal government primary responsibility for setting, implementing, and enforcing water pollution controls. As a result of the act, controls are no longer based on a case-by-case evaluation of the effects of particular pollutants on a particular stream—the "water quality" approach. Instead, uniform minimum limitations on discharges are to be set for each major polluting industry. These uniform standards eliminate many of the inequities and uncertainties of the water quality system. The nationwide program has enabled many states—including Minnesota, California, New Jersey, and Maryland—to develop and implement tough clean water programs without fear of losing industry to states with weaker programs.

The Clean Water Act has led to the clean up of some of our most polluted waters in the nation, including the St. Louis River in Minnesota, the Potomac River, and the Hudson River. Nationwide, the degradation of our waters has halted and pollution levels have remained the same despite a growing population and economy.

Most of these successes involve cleaning up conventional pollutants such as suspended solids, oil, grease, and others. Problems with toxic chemicals—the invisible killers in our water—persist. Although many waters that once were little more than open sewers are now clean enough for fish to survive, in some of these waters, the fish are contaminated by toxics and are unfit for human consumption. We still have a long way to go before we have restored our waters and met the goals of the Clean Water Act.

The Clean Water Act was designed to control pollution in two ways. In addition to the uniform effluent limitations, states are required to set "water quality standards," which establish limits on pollution for specific bodies of water, based on their optimum use. Waters designated for uses that require high quality waters—such as trout fishing—are required to be kept cleaner than waters to be used for low-quality uses, such as navigation. . . .

The EPA

The Environmental Protection Agency proposed changes in these regulations that could halt, and perhaps reverse, ten years of progress in restoring and protecting our waters. These proposals also threaten future efforts to control deadly toxic discharges.

The Izaak Walton League opposes the EPA's proposals to alter the existing regulations because they would:

• eliminate mandatory state review of all water quality standards;

• allow states to "downgrade" or weaken water quality standards more easily;

• limit the "anti-degradation policy" to the protection of existing uses only;

• allow "variances", or exemptions, for heavily polluting plants from needed pollution controls; and

• reduce opportunities for public participation in state clean water programs.

Together, these proposals signal a major retreat from meeting the goals of the Clean Water Act.

Downgrading Water Quality Standards

Nearly all of the nation's waters are designated for fishing, swimming, or recreation. Forty of the 50 states classify all of their waters for fishing or swimming. Nearly all of the waters in the other ten states are similarly classified; only a few waters in each of these states are classified for some lower quality use.

Meeting these uses will be beneficial because many people depend on fishing, boating, and other water activities for their livelihood or recreation. Additionally, meeting these uses will indicate that we are achieving our ultimate objective to "restore and maintain the chemical, physical, and biological integrity of the nation's waters". (Clean Water Act, Section 101 (a).)

The current rules require a state to maintain existing water quality standards unless it demonstrates that a stream's designated use cannot be met because the stream is hopelessly polluted, or requiring the controls needed to meet the use would cause substantial and widespread economic and social impact.

EPA's proposals would replace these simple, straightforward rules with requirements for downgrading that are vague, open-ended, and in several cases, *unrelated to water quality*. EPA's proposed regulations would allow states to permanently downgrade standards and allow more pollution if:

• abnormally low stream flows during a dry spell cause temporary reductions in fish populations;

• a stream designated for swimming is too shallow to actually swim in; or

• the benefits of meeting a standard "do not bear a reasonable relationship to the costs".

Rather than going to the trouble of requiring the clean up of fouled streams, states could instead easily downgrade standards, thereby defining away pollution problems.

EPA claims that these changes are needed because some states adopted water quality standards that were unreasonably strict and costly, and wants to allow states to weaken these standards to allow more pollution in the waters. EPA has yet to demonstrate actual instances where the existing regulations have been inflexible and have prevented a state from downgrading a standard that could not be met. In fact, under the existing regulations, Minnesota has recently downgraded 208 stream segments, covering approximately 800 miles, because the state believed that they could never become fishable or swimmable. If EPA permits this amount of downgrading of standards now, we can expect much more downgrading under these proposals. This would reward states and industries that have done a poor job of cleaning their waters. EPA's downgrading proposal makes achieving the fishable, swimmable goal of the act more difficult.

Significantly, EPA would also eliminate the requirement that states must upgrade, or strengthen, water quality standards if a stream is meeting a higher use than the standard requires. For instance, Minnesota is evaluating whether to upgrade the water quality standard of the once-poisoned St. Louis River. Eliminating the mandatory upgrading provision could put an end to this effort to ensure that current improvements in water quality become permanent.

Limiting the Anti-Degradation Policy

EPA's proposals would curtail protections for high quality waters—including outstanding national resource waters such as those in parks and wildlife refuges—that are cleaner than the strictest water quality standards requires. These waters could be polluted so long as the existing uses were still met. This change in "anti-degradation" rules would encourage states to strive for only the *minimum* level of pollution clean up needed to meet water quality standards.

Currently, only nine states have any designated outstanding national resource waters, as provided for in the existing regulations. Other states are currently developing protections for such waters in their states. This type of program ensures that these outstanding waters remain clean, and other improvements in water quality are not lost. Yet these programs could be jeopardized if EPA's regulations no longer require them. A strong federal anti-degradation policy is a cornerstone of the Act's objective to restore, protect, and maintain our nation's water quality.

Taken together, EPA's proposals would fundamentally weaken state clean water programs. The proposed changes in downgrading and anti-degradation regulations, combined with the new variance provision, will give recalcitrant industrial polluters incentives to go "pollution shopping"—closing plants in states with tough standards or locating plants in states with weak ones. Economically hard-pressed states that have strived to clean up their water could now be forced to choose between weakening their pollution controls or losing industrial jobs.

The philosophy underlying EPA's proposals is that water pollution control efforts should only strive to meet water *uses,* and not achieve water *quality.* For instance, EPA advocates limiting the anti-degradation policy for high quality waters to "the protection and maintenance of existing uses". And EPA asserts that "there is no accepted hierarchy of uses"—a poor quality use like irrigation is as desirable as a high quality use like fishing. This philosophy is grossly inconsistent with the Clean Water Act's objective to "restore and maintain" the nation's waters, and is at odds with the act's interim goal of achieving water quality suitable for fishing and swimming. The Clean Water Act is a water *quality* law. The Cuyahoga River in Ohio is now meeting its use—you can catch fish there. But the river still isn't clean—you can't eat any of the fish you catch.

The existing regulations are designed to enhance water quality by greatly restricting downgrading of standards, and by protecting high quality waters from degradation. If adopted, EPA's proposals would significantly lower the sites of the nation's clean water program. If that happened, the Clean Water Act might be more aptly called the "Clean Enough Water Act". . . .

Retain Strong Water Quality Standards

The existing regulations have proven strict enough to maintain strong state water quality standards, while remaining flexible enough to allow necessary adjustments in standards. House of Representatives bill 3282 would incorporate the key principles of the existing regulations in the act, ensuring that this delicate— yet critical—balance is maintained.

Specifically, HR 3282 would:
- allow downgrading of stream standards only when:
  —existing standards are not attainable because of "natural background" levels of pollution;
  —existing standards are not attainable because of "irretrievable, man-induced conditions" or
  —the additional cleanup beyond BAT required to meet existing standards would cause "substantial and widespread adverse economic and social impact";

- require states to upgrade water quality standards for streams that actually are cleaner than required, and are meeting a designated use that is "higher that the existing standard;
- require states to incorporate numerical toxic criteria for any of the 65 priority pollutants which reasonably could be expected to interfere with the attainment or maintenance of their water quality standards;
- prohibit the degradation of high quality waters that are cleaner than required by the most stringent water quality standards, unless they must be degraded to allow essential economic or social development for which there are no feasible alternatives. States must ensure that such degradation does not interfere with or injure existing designated uses; and
- protect "outstanding national resource water"—such as those in national parks and wildlife refuges—from degradation.

As regulations, these requirements have been a key tool in our efforts to achieve fishable, swimmable waters.

The requirement in HR 3282 that states adopt numerical water quality criteria for toxics is a new requirement, but it is consistent with and critical to achieving the goals of the act. Currently, only about half of the states have any numerical standards for toxics, and they usually cover only a handful of heavy metals. These standards rarely consider the chronic effects of these toxics or their exposure to humans through the food chain.

For instance, only 26 states have numerical standards for lead, cyanide, and cadmium. These are three of the most commonly found toxics in industrial wastewater discharged into our waters and sewers, and can cause cancer, birth defects, and other health problems in humans. Fewer than 15 states have numerical standards for PCB's. Fewer than 25 states have numerical standards for mercury, and only nine of these have standards stringent enough to protect human health, based on the EPA "White Book criteria". Many of the standards for other toxics are also considerably less stringent than EPA's national criteria for those substances. . . .

We strongly urge the Congress to stand firm in its resolve to determine the national policy for clean water. This is a national policy that should be established by Congress, and not by unelected bureaucrats at EPA or OMB.

THE STATES SHOULD LEAD THE WAY

National Association of Manufacturers

The following statement was excerpted from comments by Harold Jensen, director of energy conservation and environmental control of the Warner-Lambert Company on behalf of the National Association of Manufacturers.

Points to Consider

1. Why are water quality decisions best made at the state level?
2. How is the National Association of Manufacturers defined?
3. What are the four basic sources of water pollution?
4. When should Best Available Technology (BAT) be applied?

Excerpted from a position paper issued by the National Association of Manufacturers, April 6, 1983.

Water quality decisions are best made by the states with full public participation in the decision-making process.

The use of our nation's water is a great economic asset for our country and is indispensable to creating jobs and producing the goods that provide a higher standard of living. Therefore, our membership is greatly affected by national and state policies regarding water use. Water plays a vital role in manufacturing processes, although only a small percentage of the water used is actually consumed. Substantially more than 90 percent is returned to receiving bodies of water. The vast bulk of it is used only for cooling purposes and does not come in contact with the manufacturing process.

EPA's proposed revisions to existing water quality regulations appear to provide greater flexibility for maintaining water quality conditions consistent with practical physical limitations and economic development. Water quality decisions are best made by the states with full public participation in the decision-making process. Under present regulations, some streams are governed by water quality standards that are impossible of attainment. A greater role for the states would provide needed flexibility and recognition of local and regional realities. We are in favor of improving and maintaining stream quality conditions to the greatest extent consistent with economic and physical limitations. Thus, we believe that, once a use has been attained, it should be changed only under highly unusual circumstances, but the door should be kept open for changes when imperatively needed. It should be kept in mind that new manufacturing facilities will be using state of the art pollution abatement technology, and the stream quality will probably not be significantly diminished by siting of modern facilities in instances where industrial use would be made possible.

If industry's options for installing new facilities are unduly limited by stringent water quality policies, there will be a tendency to prolong the life of older facilities. These older facilities, even if retrofitted, will probably discharge more pollutants into the total environment than new facilities, into which the most modern pollution abatement procedures can be designed. The Congress and EPA should be continually aware that overregulation in one environmental area can lead to an overall decrease in environmental quality by increasing or prolonging pollution elsewhere . . .

52

Best at Local Level

NAMF is committed to the principle that the Clean Water Act can best be implemented at the local level where appropriate, providing site-specific flexibility and targeted controls focused on a particular local water quality situation.

National Association of Metal Finishers, 1984

NAM's Recommendations

NAM is a voluntary association of more than 12,000 enterprises engaged in manufacturing in the United States. Many of its members are substantially affected in their manufacturing procedures and in their production costs by the requirements of the Clean Water Act. Industry overall has spent billions of dollars to comply with those requirements and is continuing to do so. It is of vital importance to small business and larger business alike that the requirements of the Act be tailored so as to achieve the control of water pollution through sound and effective engineering methods.

In January of 1982, NAM's recommendations for essential changes in the Clean Water Act, a product of nearly a year's effort, were published by our Environmental Quality Committee. This publication, updated to March, 1983, discusses in some detail the reasonable and needed changes that will strengthen the Clean Water Act while enhancing our ability to meet our national water quality objectives in a more efficient and effective manner. . . .

Industrial Compliance with Clean Water Act Requirements Has Been Unsurpassed

There are four basic and distinct sources of water pollution in this country: (1) natural, "background" pollutants that appear in our rivers and streams *not* as a result of human activity; (2) "non-point sources" of water pollution that are a result of runoff from agricultural, urban street wash, construction, silvicultural and other activities; (3) publicly-owned treatment works (POTWs) that treat wastewater from industrial, residential, commercial, and institutional sources and then discharge the treated effluent; and (4) industrial sources that treat their own wastewater before discharging it. Introduction of natural pollutants cannot be controlled and this natural background should be

Source: EPA

taken into consideration in any control program. Non-point sources have long been recognized as being a major contributor to water pollution, but have not been effectively addressed. POTWs have a rather spotty performance and Congress has rightfully begun to take steps to improve it. Of the four sources, only direct industrial dischargers have done the job required of them by Clean Water Act regulations.

EPA has reported that in the case of POTWs, for instance, at any given point in time 50 to 75 per cent of the plants are in violation of their permits. A November 1980 General Accounting Office Report surveyed 242 POTW plants in 10 states, and found that 87 per cent were in violation of their permit and 31 per cent, in GAO's opinion, were in serious violation. Direct industrial dischargers, on the other hand, report 90 – 95 per cent plus permit condition compliance rate. From this, it is obvious that the most hopeful areas for continued dramatic improvement in water quality lie not with direct industrial dischargers, but with POTWs. The following is an extract of some of our major recommendations:

National Pollutant Discharge Elimination System (NPDES)

—Uniform national categorical pretreatment standards should be eliminated and statutory compliance repealed. POTW authorities should have the responsibility and authority to implement their own pretreatment programs which would include the establishment of local pre-treatment standards consistent with the needs of the POTW for NPDES permit compliance. EPA approval of local programs should not be required. All POTW NPDES permits should be enforced in the same manner as industrial NPDES permits, placing the responsibility for POTW discharges on the municipality. . . .

Best Available Technology Economically Achievable (BAT)

—BAT limitations should be required only where there is a significant toxics problem. "Significant toxics problems" should be defined as those where BPT (Best Practicable Technology) standards or limitations are not providing sufficient protection to the receiving waters and where further abatement would have a *measurable* effect on receiving waters.

—Situations where a pollutant is present in the effluent solely as a result of its presence in intake waters should not be considered a significant toxics problem. The legislation should provide flexibility so that States and localities can administer reasonable programs.

—BPT standards should not be elevated so as to equal BAT or require zero discharge.

—BAT requirements for nonconventional pollutants should not be applied unless water quality demands it. . . .

Conclusion

The Congress has an opportunity in the course of reauthorizing the Clean Water Act to make a number of changes that would facilitate the achievement of better water quality in this country. The changes we have suggested would simplify and streamline the law, giving it greater flexibility and creating a true Federal-State partnership.

PUBLIC FAVORS STRONG CLEAN WATER ACT

Louis Harris

The following comments by Louis Harris describe the results of his firm's study of public attitudes toward water pollution control and conservation of water resources conducted for the Natural Resources Council of America. This polling survey was conducted among a cross-section of 1,253 adults nationwide.

Points to Consider

1. What did the Harris survey say about national attitudes toward strong government actions to curb air and water pollution?
2. How do people feel about the Clean Water Act?
3. What are the "razor's edge issues" and how are national attitudes toward these issues described?
4. How are public attitudes toward the costs of clean water described?

Excerpted from testimony by Louis Harris before the Senate Committee on Environment and Public Works, December 15, 1982.

The American people do not feel they must opt for economic growth, on the one hand, or an enviromental clean-up, on the other.

One of the most common claims one hears is that while it is desirable to clean up the environment, the Public gives such efforts at best a secondary priority. In times of unemployment, the plea is that an environmental clean-up must wait until there are enough jobs, or the clean-up ought to be delayed or eased to preserve jobs. In times of energy shortages the efforts to clean up the environment are asked to wait once again until enough abundant energy is available.

Well, in this study, we have some definitive light to shed on such claims. Curbing water pollution is named by a substantial 74% of the adult population as "very important" in making the quality of life "better". This is two points above a comparable 72% who said the same about curbing air pollution. While a bit lower than the 83% who give high importance to curbing inflation and the 79% who say the same about bringing interest rates down, curbing water pollution is viewed as equal in importance to economic growth, also cited by 74% as very important, and outscores the 65% who felt that way about producing more energy, and the 68% who said the same about making products and services safer.

Economic Growth

The inescapable point is that curbing both air and water pollution are given a high priority of importance in their own right by the American people. The basic truth, however, is that by a sizable margin, the American people do not feel they must opt for economic growth, on the one hand, or an environmental clean-up, on the other. Or, of finding jobs to ease unemployment, on the one hand, and an environmental clean-up on the other. Back in 1975, we found that by 64–22%, a solid majority of the public then rejected the notion that it was "necessary to slow down the rate at which we clean up water pollution in order to get the economy moving again and to ease unemployment." Now, some 7 years later, in 1982, sentiment has grown even more overwhelming: By 89–6%, almost all Americans are convinced that such a trade off is not necessary, because they are firmly convinced that we can "get the economy going again and also clean up water pollution." . . .

Indeed, when asked what has happened to their own streams, lakes and rivers over the past 10 years, 46% nationwide report

> **People want to take absolutely no chances with their drinking water where human health and safety are concerned.**

where they live they are *more* polluted now, only 17% say less, and 30% the same. And, looking forward to the next 10 years, 44% feel their own streams, lakes, and rivers will be *more* polluted, only 17% feel they will be less polluted, while 34% anticipate no real change. Such results underscore the concern the American people have that their basic water resources are threatened by pollution and constant and tough measures must be taken to keep them clean or to clean them up.

Clean Water Act

Significant shifts have taken place just this year in basic attitudes about extension of the Clean Water Act. Last year, only a very small 6% of the public nationwide felt that federal water pollution standards were overly protective. Now, an even smaller 4% feel that way. A year ago, 43% felt that federal water pollution standards were "just about right", but the number who feel that way now has dropped to 34%. The big movement has been on the dimension "not protective enough", which has jumped from 48 to 59%, just in one year. I believe the clear implication from this shift is not simply that the number who reject the claim that the federal law is overly protective has gone from 91–6% to 93–4%, but that there is growing and deep concern that while the standards may be very good, indeed, the enforcement of those standards has deteriorated, as control of the federal establishment has fallen into hands less than fully in accord with the letter and spirit of the Clean Water Act of 1972.

When we ask directly about the Clean Water Act, a 94–3% nearly unanimous majority wants it kept at least as tough and indeed, fully 60% want to make it "stricter." Last year, a comparable 93–4% majority shared the same view, but the number who wanted the law more strict was a lower 52%. Thus, the sentiment for making Clean Water tougher rather than easing it has risen from 52 to 60% in a year. Nor are people in a temporizing mood on environmental matters. Even though not many experts think that all the nation's waters will be made "fishable" and "swimmable" by 1983, as set forth in the original Clean Water Act, by 78–17%, a big majority are convinced that the job can be done. The public has both the faith that the task can be accomplished and is deeply committed to getting the job done.

58

Much of our leisure time depends on clean water: boating, swimming, fishing, and sunbathing.
Source: Water Pollution Control Federation

Thus, make no mistake about it, any effort to weaken or to ease the Clean Water Act will be viewed by the American people as a serious and inexcusable violation of a clearly expressed and deeply felt public mandate.

Razor's Edge

Now, let's get down to some of the razor's edge issues that are the heart of the battle over Clean Water today. One of the 1977 amendments to the Clean Water Act, as you know, specifies that factories must install the best available pollution control technology (BAT) by 1984. Some industries would like to change the Act so that the government could grant waivers or variances from the BAT requirement under certain circumstances. We asked a question about this matter in a way that bent far over backwards to get the full weight of the opposition to BAT into the query. We asked: "The Clean Water Act now requires factories to install the latest, most effective systems available to clean up their toxic pollutants. *Given the high cost of meeting this requirement,* which *one* of the following policies do you think the government should follow—require all factories to install the best anti-pollution systems they say they can afford, or require all factories to install the best anti-pollution system available, except those factories which would be forced to shut down as a result, or require all factories to install the best anti-pollution systems available," the last obviously without any exceptions or exemptions based on cost. Well, no more than 21% opted for installing BAT on the basis of what they can afford, and 23% more opted for BAT, except when it would make the plant close down. A solid 53% majority said it should be mandatory that BAT be installed, regardless of cost or consequences.

We asked much the same type of question, when we queried the cross-section: "Do you think factories should be required to install the best anti-pollution systems available, even if this meant that fewer jobs would be available or not?" By 65–27% nationwide, a big majority said put in BAT, even if that means fewer jobs. All groups favor BAT by margins of 2 to 1 or better. In the West, they support it by 75–18%, among union members a 65–28% majority favor BAT, and skilled labor opts for it by 73–21%.

We then asked another trade-off, offering people a choice between waiving BAT for toxic pollutants or causing the company involved to lose money, shut down the factory, and lay off workers. By an overwhelming 86–12%, people said that the waiver or variance should not be granted.

Costs of Clean Water

In this survey, we probed in considerable depth on the issue of cost-benefit analysis, another of the key and pivotal issues at stake in the Clean Water debate. We asked: "Some people feel that the government should not order industries to clean up their water pollution unless it can be shown that the benefits of the cleanup justify the costs involved. Others disagree and say the value of eliminating unsafe pollutants in water can't be measured in dollars. Would you favor changing the Clean Water Act so that the government couldn't order an industry to clean up its water pollution *if the costs didn't justify it,* or would you oppose such a change in the Clean Water Act?" By a striking 74–23%, a solid majority say they would oppose changes even if the cost didn't justify it. Obviously, people mean business on the subject of cleaning up water pollution.

Or take the case of safe drinking water. By 92–7%, a big majority of the public reports that drinking water that comes out of their tap or faucet in their home is either "very" or "pretty" safe. Despite this confidence, nonetheless, by 65–33% a solid 2 to 1 majority of the public want to follow a policy in the renewal of the Safe Drinking Water Act that is also up for reauthorization which states "contaminants in drinking water supplies ought to be cleaned up if there is *any chance* that they will be dangerous to human health, regardless of the cost." Even though they think their drinking water supply is safe, people want to take absolutely no chances with their drinking water where human health and safety are concerned. Any provision that requires proof of "clear danger" will meet with stiff public resistance. . . .

Conclusion

Those in charge of running environmental programs in Washington today are widely viewed as wanting to ease strict enforcement of federal environmental controls. It is evident that the American people want strict environmental controls, but also are fearful that those now vested with responsibility of setting and enforcing standards are quite dedicated to doing just the opposite. In turn, this has mooted and clouded any clear-cut choice in the federal vs. state and local control issue. Obviously, the public would like to have continuation of federal controls, but is worried that those now in authority will not enforce either the spirit or the letter of the law.

As the congressional process deals with renewal of the Clean Water Act, an alert public will be drawing a judgment not only on how well the basic provisions of the act are kept and strengthened, but also what safeguards are included to make certain strict standards are enforced.

NEW THREATS FROM TOXIC WASTES

The Sierra Club

Pamela Brodie wrote the following article for the Sierra Club as part of a pamphlet titled The Clean Water Act: New Threats from Toxic Wastes Demand Stronger Law. *She wrote this statement in her capacity as a Sierra Club lobbyist for the Clean Water Act.*

Points to Consider

1. What progress has been made in curbing water pollution?
2. How has the Reagan Administration responded to the issue of water quality?
3. What distinction is made between conventional and toxic pollutants?
4. How are (BPT) and (BAT) control technologies described and defined?
5. What policies toward water quality standards are recommended?

Excerpted from a public position paper issued by the Sierra Club, September/October, 1983.

Much of the controversy over the reauthorization of the Clean Water Act surrounds attempts to weaken the act's controls over toxic substances.

In the 1950s and 1960s the Cuyahoga River ran a muddy brown through Akron and Cleveland, Ohio. Steel and chemical factories along its lower course discharged some 155 tons a day of wastes associated with the production of chemicals, oil, and iron. Raw or inadequately treated municipal sewage added to the problem; some days the bacterial count in the river was as high as in a sewer. Miasmic gas from decaying organic matter bubbled to the surface. In 1959, the Cuyahoga River caught fire and burned for eight days. A decade later, it burst into flames once more.

In 1972, public outcry against water pollution forced Congress to take action. It amended the Federal Water Pollution Control Act of 1956, and for the first time the federal government assumed the lead in regulating water pollution. Starting in 1972, uniform nationwide controls were established for each category of major polluting industry.

In 1977 the law was reauthorized and its name changed to the Clean Water Act. The objective of the act was "to restore and maintain the chemical, physical, and biological integrity of the nation's waters." At the same time, the federal government agreed to pay the lion's share of the cost of new sewage-treatment plants. . . .

Many rivers, streams, and lakes that were practically dead a decade ago are now thriving. The Hudson River supports striped bass again, and the Potomac River is no longer choked with algae. The Cuyahoga is now the center of a popular national recreation area. . . .

The Reagan administration launched a three-pronged attack on the Clean Water Act:

• *Budget cuts.* The administration reduced by 40 percent the 1982 and 1983 funding for the Environmental Protection Agency, which enforces the Clean Water Act.

• *Regulations.* The EPA under the Reagan administration has proposed changes in regulations that would allow states to more easily weaken water-quality standards that require the eventual attainment of "fishable-swimmable" waters.

• *Legislation.* Administration-backed proposals would delay or weaken controls on toxic discharges by industry for several years beyond the time companies actually need to install necessary equipment.

Conventional Pollutants and Toxic Substances

One of the initial focuses of the Clean Water Act was "conventional" pollution—human and animal wastes, eroded soils, and organic debris. Most sewage-treatment plants were built and testing methods devised only for these pollutants. Since 1972 the federal government has spent more than $37 billion on sewage-treatment plants. Some 4,500 new plants have been completed, and 8,200 more are under construction. As they are completed, conventional water pollution should decrease.

Nevertheless, many communities still will not have adequate sewage-treatment facilities. In 1980 the EPA estimated that more than 60 percent of the nation's municipal treatment plants did not meet the minimum goals that were to have been met by 1977. Furthermore, many of the nation's older treatment systems are in need of replacement; this is part of the "infrastructure" crisis that threatens America's sewage systems along with its highways, bridges, and transportation systems.

Industries have spent more than $30 billion to comply with the Clean Water Act and have achieved impressive results cleaning up conventional pollutants, such as wastes from pulp and paper mills and from food-processing plants. Another threat to rivers and lakes has only recently been widely recognized: toxic wastes.

According to Sen. Robert Stafford (R-Vt.), chair of the Senate Committee on Environment and Public Works, "persistent toxics constitute the key problem in our current efforts to protect the quality of our nation's waters." These new types of pollutants, dangerous in relatively small quantities, come in a bewilderingly large (and growing) variety of forms that are sometimes hard to detect. Many toxic substances do not readily decompose. Some such substances, including DDT, mercury, lead, and polychlorinated biphenyls (PCBs), concentrate as they move up the food chain; concentrations of dangerous chemicals in fish are thus thousands of times greater than those in surrounding waters. Concentrations increase even further in humans who eat the fish, so that commercial and recreational fishing is restricted in many areas. For instance, Michigan's state fishing licenses contain this warning: "Certain Great Lakes fish should not be eaten—by children, women who are pregnant, nursing, or expect to bear children. Limit consumption by all others to no more than one meal per week." . . .

Much of the controversy over the reauthorization of the Clean Water Act surrounds attempts to weaken the act's controls over

64

toxic substances. The act aims to eliminate toxic discharges by encouraging industries to reclaim and reuse chemicals rather than discharge them. More than 320 million pounds of toxic organic chemicals are discharged each year into the nation's waterways and sewers; 80 percent of them come from the organic-chemicals industry. . . .

Direct Discharges

One of the most controversial questions concerns methods of controlling toxic discharges. Currently, industries that discharge pollutants directly into streams must employ two levels of control. The minimum level is called "best practical control technology" (BPT), which 90 percent of the nation's industries have attained. BPT is designed to remove only conventional wastes, though it does incidentally remove between 50 and 70 percent of the toxics.

The next step is "best available" technology economically achievable" (BAT), which would remove most of the remaining toxics. The EPA is establishing BPT and BAT guidelines for 25 categories of major polluting industries, including electroplating, organic chemicals, steel, and pulp and paper manufacturers. The stringency of the BAT guidelines is a matter of intense concern. In some cases the Reagan EPA has issued BAT guidelines that are no stricter than those for BPT. However, the organic-chemical industry and other must meet BAT guidelines that are much stricter than BPT, since these industries are the major sources of toxic pollution.

The Clean Water Act now requires industries to comply with BAT guidelines by July 1984. Since the EPA has not yet released all the BAT guidelines, it is impossible for most companies to comply with the rules in time. The EPA and some industry representatives have testified that two or three years is enough time to comply with regulations. Nevertheless, some industry lobbyists are working for longer delays or waivers of the requirements. Conservationists have been urging that compliance be delayed no longer than three years from the time a regulation is issued.

Water Quality Standards

A second area of concern in the debate over reauthorization of the Clean Water Act is water-quality standards. The 1972 act required states to classify rivers and streams for various uses. One of the common standards of water quality—a goal to be achieved—is "fishable-swimmable," which means that the water should be clean enough for fishing or swimming.

In October 1982 the Reagan EPA proposed changes in regula-

400 Million Pounds of Toxics

Recent EPA data suggests that American industry is discharging over 400 million pounds of toxics into US waterways each year—a very conservative estimate. Contaminated fish, one symptom of this problem, have become so common that many states warn fishermen to eat little, if any, of their catch.

Engage/Social Action, 1984

tions that would allow individual states to loosen water-quality standards more easily, retreating from the "fishable-swimmable" standard. This policy ostensibly returns the duty of regulation to state authorities, but it also opens the door to a sort of economic blackmail. States could be tempted to trade pollution control for economic development, health for jobs. The proposed regulations would also eliminate the national policy of protecting pristine waters in parks, wilderness areas, and wildlife refuges.

In some areas with high volumes of industrial wastewater, waters will not be clean enough to meet "fishable-swimmable" standards even after BAT controls are in place. Conservationists want the Clean Water Act amended so that the EPA would be required to identify such "toxic hotspots" and to develop and implement additional controls over toxic discharges in order to clean these areas up.

Pretreatment

While BAT regulations apply only to companies that discharge wastes into streams and other open waterways, more than half the nation's industrial polluters (about 60,000 facilities) discharge their wastes into municipal sewage systems.

Keeping toxic discharges out of sewers is essential for a number of reasons. Sewage plants are designed mainly to process domestic sewage. The main byproduct of such plants is sludge. Uncontaminated sludge—sludge without toxic wastes—can be recycled as fertilizer. Sewage sludge contaminated by toxic substances must be disposed of as a hazardous waste—buried in landfills, incinerated, or dumped in the oceans, all practices potentially harmful to the environment. Such large quantities of diluted toxic wastes are much more expensive to dispose of than are the small residues that are removed at the source by pretreatment.

Moreover, toxic substances can in fact directly interfere with the workings of a sewage-treatment plant. The toxic substances can kill off the bacteria that are used to digest the conventional wastes, rendering its treatment processes useless. Finally, toxics can pass through the plant untreated and be discharged into receiving waters. . . .

The only way to prevent damage to sewage plants from toxic substances is to keep the two separate, a goal that requires pretreatment of toxic wastes. However, the federal government's program of pretreatment is under attack. Some industries and municipalities are urging that individual communities be allowed to "opt out" of the federal program by providing their own pretreatment programs. In other words, some industries argue that individual communities should be permitted to set their own standards.

Conservationists oppose the proposal because it would allow greater discharges of toxic substances into the environment. Though many communities already have sewage-treatment plants that can remove toxic substances, such facilities are often far from adequate. At present, according to Daniel Weiss of the Izaak Walton League of America, sewage-treatment plants with pretreatment programs are "designed to do the minimum—like protect the plants from exploding—but not really to ensure that toxics aren't passing through the plant and into the receiving waters."

THE SAFE DRINKING WATER ACT

Natural Resources Defense Council

The Natural Resources Defense Council (NRDC) is a national non-profit environmental organization supported by 40,000 members and contributors dedicated to protecting public health and preserving the quality of the natural environment. Since its inception in 1970, NRDC's staff scientists and attorneys concerned with toxic substances control have sought, through legislative, administrative and judicial proceedings, to achieve a reduction in involuntary human exposures to carcinogenic and other toxic environmental contaminants. With respect to water quality, these efforts have focused upon obtaining controls on pollution at the source, under the Clean Water Act and the Resource Conservation and Recovery Act (RCRA), and at the tap, under the Safe Drinking Water Act (SDWA).

Points to Consider

1. How many chemicals have been detected in the nation's drinking water?
2. What percent of groundwater drinking water supplies have been contaminated?
3. Why is the EPA record for implementing the Safe Drinking Water Act deserving of criticism?
4. In what ways should the Safe Drinking Water Act be written to adequately protect the public?

Excerpted from a public statement on the Safe Drinking Water Act by the Natural Resources Defense Council, June, 1984.

As a matter of policy, the Act favors protecting public health rather than placing the public at continuing risk.

Congress passed the Safe Drinking Water Act (SDWA) at the end of 1974, in response to a widespread national controversy over the extensive presence of organic chemicals, many of which are carcinogenic, in drinking water supplies throughout the United States. Since then, the evidence that our water supplies are contaminated by dangerous organic and inorganic chemicals has continued to mount. Organic chemicals such as trichloroethylene, tetrachloroethylene, vinyl chloride, carbon tetrachloride, aldicarb, EDB, chloroform, and benzene have been frequently detected in drinking water.

Although the National Academy of Sciences has identified as carcinogens 22 of the chemicals found in drinking water, and many others are known to be mutagenic or otherwise toxic, the vast majority of the estimated 700 chemicals so far detected in drinking water have not yet been tested to determine their toxicity. Moreover, we now know that a significant portion of our groundwater supplies, upon which half the U.S. population depends for drinking water, are contaminated with industrial chemicals and pesticides. According to a 1982 report by the Office of Technology Assessment (OTA), 29% of the groundwater drinking water supplies of 954 United States cities with populations over 10,000 are contaminated. Similarly, in a Federal Register notice published last year, EPA announced that organic chemicals have been detected in 45% of the public water systems that draw on groundwater and serve over 10,000 people.

The known sources of contaminants in groundwater include disposal of hazardous wastes in landfills and industrial surface impoundments (pits, settling ponds and lagoons); septic tanks and cesspools and the chemicals used to clean them; municipal wastewater; mining activity; petroleum exploration and development; underground injection of wastes, and agricultural, street and urban runoff. Because groundwater lacks the self-cleansing properties provided to surface water by dilution, circulation, and degradation by sunlight and aquatic organisms, groundwater can remain contaminated for centuries. Furthermore, concentrations of contaminants in groundwater can be orders of magnitude greater than levels found in surface water. For example, in 1981 the Council on Environmental Quality (CEQ) reported that the highest level of the widely-used industrial solvent trichloroethylene measured by then in surface water was 160 ppb (parts per billion). By contrast, the highest level so far found in groundwater is 35,000 ppb.

Preventive Approach

In light of the potential human hazards posed by exposure to a broad array of dangerous chemicals in drinking water, Congress adopted a preventive approach in the SDWA which requires that uncertainties be resolved on the side of protecting public health. As a matter of policy, the Act favors protecting public health rather than placing the public at continuing risk. The alternative, waiting until the link between disease and consumption of chemically-contaminated drinking water is conclusively proved, is contrary to accepted preventive health philosophy and could impose an unnecessary and unconscionable social cost in terms of future human pain and suffering.

The Environmental Protection Agency's record in implementing the SDWA, and especially the Agency's continuing failure to establish national primary drinking water regulations for organic chemicals, is deserving of strong criticism. Since 1975, EPA has issued only a very short list of maximum contaminant levels (MCLs). The list includes a few pesticides and herbicides, a small number of inorganic chemicals, a standard for coliform bacteria, turbidity, radionuclides, and, since 1979, trihalomethanes. . . .

EPA's Failure

The consequences of EPA's failure to establish standards for organic chemical contaminants in drinking water are very serious. First, because the establishment of standards is a necessary prerequisite for triggering monitoring of drinking water quality, water suppliers are not presently required to monitor the water they sell for organic chemical contamination. The only contaminants they must look for are the relatively small number presently included in the Interim Regulations. Thus, unless a supplier voluntarily chooses to do so (as have a number of the nation's larger water companies) or else chemical contamination of a supply is so gross as to cause taste, odor or public health problems, most water utilities across the country do not routinely monitor for unregulated contaminants and, accordingly, do not know the chemical quality of the water they provide.

Second, and equally important, other regulatory programs designed to prevent contamination of groundwater are keyed to the National Primary Drinking Water Regulations. Under the Resource Conservation and Recovery Act (RCRA), leachate from hazardous waste landfills is only tested for the presence of chemicals included on the Agency's short list of primary drinking water standards. Since most of the chemical groundwater contaminants believed to have resulted from the burial of hazardous waste are *not* on the list, reliance on the Primary Drinking Water Regulations to protect groundwater against leaking landfills is misplaced. To illustrate, a landfill leachate containing trichloroethylene and tetrachloroethylene, two of the most widely detected groundwater contaminants, would *not* necessarily be tested for those compounds because they are not on EPA's current list of drinking water standards. Thus, landfill leachate could successfully meet the requirements of RCRA although it was heavily contaminated with dangerous chemicals.

Underground Injection

The Underground Injection Control (UIC) program, established in the SDWA to prevent groundwater contamination by underground injection wells, is also keyed to the Primary Drinking Water Regulations. Under EPA's UIC regulations, migration of fluids from injection wells into underground sources of drinking water is permitted unless "The presence of the contaminant *may cause a violation of any primary drinking water regulation* or may otherwise adversely affect the health of persons." This provision permits the leaching of injected liquid wastes into underground drinking water supplies until monitoring detects contamination levels that either violate one of the short list of MCLs so

Property Values Plummet

Jeff and Marilyn Downs got an unwelcome surprise shortly after they built a $60,000, two-bedroom home in 1981 in Lake Elmo, about a mile south of Lake Jane.

Water from their new $2,500 well was found to be contaminated by industrial solvents. . . . Property values have plummeted, for their home and 38 others near the landfill. Thirty-five area landowners have had their market values reduced by more than $1 million last year, said Lake Elmo Administrator Patrick Klaers.

Minneapolis Star and Tribune, April 11, 1985.

far put forth by EPA, or else pose an actual threat to human health. In our view, this approach permits significant contamination of groundwater to occur before corrective action is required, and thus directly changes the preventive intent of the SDWA. Unless prohibitions on hazardous waste injection and expanded groundwater monitoring requirements are enacted, the potential for groundwater contamination from underground injection activities will remain.

Almost 60 percent of the liquid hazardous waste disposed of every year in the United States is underground injected. This staggering volume of hazardous waste, estimated to be 8.6 billion gallons, must be viewed as a danger to groundwater unless a convincing case to the contrary can be made.

To remedy the problems discussed above, NRDC advocates passage of legislation that would state as a national goal the assurance to the people of the United States of a safe supply of drinking water and the protection of our groundwater resources against contamination. Such legislation should also mandate the establishment of (1) additional drinking water standards to protect public health against contaminated drinking water; (2) expanded monitoring requirements for unregulated contaminants to facilitate accurate assessment of the present chemical quality of our drinking water; (3) restrictions on underground injection of hazardous waste, and other land disposal practices that endanger groundwater, and (4) requirements for development and implementation of groundwater protection plans. . . . The history of drinking water quality in this country teaches that a strong federal presence is a necessary prerequisite.

National Drinking Water Regulations Standard-setting Requirements

NRDC supports the imposition upon EPA of mandatory requirements to establish maximum contaminant levels or treatment techniques for organic chemicals and other toxic contaminants of drinking water, to be met within prescribed deadlines. Left entirely to its own discretion, the Agency has demonstrated its continuing inability to accomplish the task of identifying and setting health protective standards for contaminants of concern in the nation's drinking water supplies.

Almost ten years have passed since the SDWA became law, but after complying in 1975 with the statutory directive to codify the 1962 Public Health Service drinking water standards as Interim Primary Drinking Water Regulations, EPA has set only two MCLs, for radionuclides (1976) and for the products of chlorination, trihalomethanes (1979), during the next nine years. Instead of a comprehensive set of drinking water regulations which would serve to assure water consumers that their drinking water is safe to drink, EPA's list is notable for the widely-detected contaminants that are not on it. . . .

Conclusion

Drinking water has been a neglected resource for years and has suffered a decline in quality across the country as a result. NRDC strongly supports a well-conceived effort to strengthen the Safe Drinking Water Act by establishing a national health protection goal for drinking water, and by requiring EPA to set new drinking water standards, to initiate a monitoring program to identify contaminants that should be regulated, to curb injection practices that threaten groundwater, and generally to expand federal and state protection of drinking water quality.

Renewed Congressional concern about drinking water is appropriate in light of continuing disclosures of serious contamination problems.

FLEXIBILITY NEEDED IN WATER REGULATION

Chemical Manufacturers Association

The Chemical Manufacturers Association is a nonprofit trade association whose company members represent more than 90 percent of the production capacity of basic industrial chemicals in the nation. CMA member industries and companies are directly affected by the requirements of the Clean Water Act (CWA).

Points to Consider

1. Why are (BAT) best available technology controls on industrial polluters no longer necessary?
2. What major advances have been made in curbing water pollution?
3. How much money have companies invested in pollution control equipment?
4. When should downgrading of water quality standards be permitted?

Reprinted from a position paper issued by the Chemical Manufacturers Association on April 14, 1983.

74

CMA believes that the success achieved to date in improving water quality can best be maintained and enhanced if additional flexibility is allowed in the Act.

The Chemical Manufacturers Association (CMA) believes that the country has made considerable progress in water pollution control through implementation of the Clean Water Act in a reasonable and flexible manner. In order to continue this progress, CMA urges Congress to examine how the Act can be improved by directing attention to significant problems and ensuring that the Act addresses them in a sound manner. CMA has the following specific suggestions:

Modification of BAT limitations—The treatment technology installed by industry to date has resulted in substantial improvements in our nation's waters. Consequently, uniform implementation of BAT controls is not necessary. Modifications of BAT limitations should be allowed where an adequate environmental study shows that BAT is not necessary to achieve good water quality. Modifications will avoid costly expenditures by industry for treatment technology that will not improve environmental quality. We believe that a workable modification scheme can be established consistent with the nation's commitment to the control of toxic pollutants.

Pretreatment—The pretreatment program has centered on national technology-based standards that do not properly take into account the need for such treatment. The result is costly, redundant treatment by both industry and POTWs. CMA supports a legislative amendment to allow industrial users of publicly owned treatment works (POTWs) which establish alternative local pretreatment requirements to be exempt from the national categorical pretreatment standards. CMA has several specific suggestions for amendments to S.431 that will help the bill serve its intended purpose. . . .

Award of Fees—CMA sees no need to amend the Act to allow for the award of costs and attorney fees. The Equal Access to Justice Act, enacted in 1980, provides adequate relief to organizations or individuals that could otherwise not afford the costs of litigation. . . .

Federal Common Law—Efforts to amend the Clean Water Act to authorize suits against dischargers for abatement or damages based on federal common law, rather than on the specific provisions of the Clean Water Act, should be rejected. The courts are

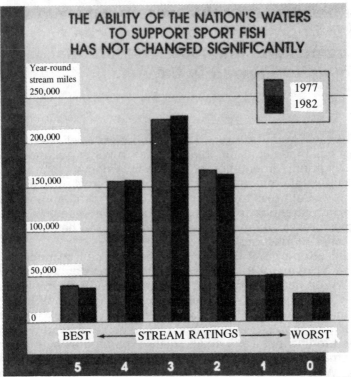

THE ABILITY OF THE NATION'S WATERS
TO SUPPORT SPORT FISH
HAS NOT CHANGED SIGNIFICANTLY

Year-round stream miles

1977
1982

BEST ◄——— STREAM RATINGS ———► WORST

5 4 3 2 1 0

Source: National Fisheries Survey. EPA and the U.S. Fish and Wildlife
Service, 1984.

less capable than the Congress or EPA of developing and implementing national environmental policy. . . .

The installation of pollution control equipment by American
industry, in many cases prior to the enactment of the Clean
Water Act, has brought about major advances in cleaning up
and protecting the nation's waters. I am speaking here not only
of conventional pollutants such as suspended solids or pH, but
also of the toxic or priority pollutants that have been a specific
concern since the 1977 amendments. As we reported a study
completed last year on joint CMA/EPA analyses of samples from
industry wastewater treatment systems convincingly demonstrates that widely used biological treatment systems, which EPA
has identified as the best practicable technology (BPT) for the organic chemicals industry category, remove a high percentage of
most organic priority pollutants.

We believe that much of the improvement in water quality
that the country has witnessed results from the flexible and reasonable manner in which EPA and the states have implemented
the Clean Water Act through the NPDES permit system. CMA

76

urges the Congress to evaluate carefully the progress that industry has made during the past decade in water pollution control and to assess to what extent additional controls on industrial discharges will be necessary or effective to achieve further improvements in water quality.

Current policies under the Act should be examined to ensure that future requirements are imposed in a flexible manner to allow alternative treatment to be used where it will protect the environment. This will avoid devoting industrial capital to expensive treatment systems that will not provide water quality improvements. These principles apply with particular force to the BAT and pretreatment provisions of the Act.

Modification of BAT Limitations

A central feature of the Act's control of the discharge of pollutants is the requirement that industrial dischargers to the nation's waters meet a series of increasingly stringent levels of effluent limitations. The 1972 amendments effected a shift from exclusive reliance on water quality controls to technology-based effluent limitations. This reliance on technology may have been sound in 1972. However, after a decade of experience, the time has come to evaluate whether the further application of technology-based effluent limitations will continue to improve water quality or instead will result in treatment for treatment's sake with no appreciable benefit to the quality of the nation's waters.

The Progress Achieved to Date

In the absence of effluent limitations guidelines, most of the first level of technology-based permits, namely BPT permits, were developed on a case-by-case basis using the permit writer's best professional judgment. The record using this process has been a good one. For example, by 1976, the chemical industry had dedicated $1.8 billion for equipment, which cost almost $400 million a year to run. At the end of 1981, America's chemical companies had more than doubled their capital investment in water pollution control to $3.7 billion, and in 1981 it cost $1 billion to operate those facilities. In total, the chemical industry expended nearly $10 billion for the construction, operation and maintenance of water pollution control facilities through 1981.

The treatment already installed by the chemical industry has achieved better than expected results in controlling priority (toxic) pollutants. For example, the CMA/EPA Five-Plant Study completed last year demonstrates that well-designed and well-

> **The installation of pollution control equipment by American industry, in many cases prior to the enactment of the Clean Water Act, has brought about major advances in cleaning up and protecting the nation's waters.**

operated biological treatment systems are effective in removing organic priority pollutants. . . .

EPA described the results from industrial controls under the Clean Water Act to this Committee on February 5, 1982. "From 1972 to 1977 implementation of best practicable technology regulations . . . significantly reduced industrial discharges of six key pollutants: biochemical oxygen demand (BOD) by 69 percent; suspended solids by 80 percent; oil and grease by 71 percent; dissolved solids by 52 percent; phosphate by 74 percent and heavy metals by 75 percent."

The Need for Flexibility

EPA is now about to begin the implementation of the second level of technology, namely BAT controls. CMA believes that the success achieved to date in improving water quality can best be maintained and enhanced if additional flexibility is allowed in the Act. Considering the substantial anticipated costs of across-the-board BAT controls, the uniform requirement of such controls is no longer appropriate in view of the unanticipated significant reductions in discharges of priority pollutants to date. In order to avoid the imposition of costly treatment requirements that are not necessary to achieve good water quality, we recommend that Congress amend the Clean Water Act to allow additional flexibility in the use of BAT effluent limitations. Where water quality circumstances permit, a discharger should be allowed to qualify for alternative BAT effluent limitations. In particular, the Act should be amended to allow for a modification of BAT effluent limitations where the applicant can show that the resulting discharge of pollutants will not pose an unacceptable risk of harm in the receiving water based on the water's designated use.

In contrast, the costs of meeting anticipated BAT requirements would be large. We estimate that Du Pont would have to spend at least $5 million in new capital and $2 million in annual operating costs to meet BAT requirements at this plant, based on activated carbon treatment of isolated streams. The foregoing study provides evidence that such expenditures are not warranted

78

since further treatment of Du Pont's discharge would not appear to result in any perceptible improvement in water quality. We think it would be a wise policy to allow industry the opportunity to persuade the permit authority that alternate limitations are warranted for such plants.

Water Quality Standards

EPA's current regulations governing water quality standards were adopted in May 1979. These regulations reflect a commitment to anti-degradation but also allow some flexibility where existing uses cannot be attained. The regulations require that water quality standards be established which will result in achievement of the national water quality goal wherever attainable. . . .

The requirement that designated uses existing as of January 1, 1983, be maintained at a minimum would have the effect of invalidating the current EPA regulations that allow for limited downgrading where existing uses cannot be attained due to natural conditions. We urge that Congress not revoke this sound regulatory provision, which simply allows the states to recognize when they have required the impossible. It is one thing to prevent degradation of water quality that has actually been achieved; it is quite another matter inflexibly to adhere to a standard that is impossible to attain. . . .

In summary, CMA opposes the prohibition of downgrading where existing uses cannot be attained, and sees no reason for legislation to amend the water quality provisions of the Act at this time.

BLUEPRINTS FOR CLEAN WATER

PRETREATMENT OF INDUSTRIAL WASTEWATER

Synthetic Organic Chemical Manufacturers Association

The Synthetic Organic Chemical Manufacturers Association (SOCMA) is a nonprofit association of producers of organic chemicals. Most of SOCMA's 100 members operate small chemical plants that discharge wastes into publicly owned treatment works, rather than directly into rivers and lakes.

Points to Consider

1. Why do SOCMA's members discharge wastes into publicly owned treatment systems?
2. What is the difference between direct and indirect waste discharges?
3. What kind of pretreatment strategy does SOCMA present?
4. What specific proposals are made?

Reprinted from a public statement by the Synthetic Organic Chemical Manufacturers Association, 1983.

Many municipalities already have developed effective local pretreatment programs. SOCMA believes that such programs have substantially implemented the objectives of the Clean Water Act.

The regulatory approach to pretreatment of industrial wastewater by users of publicly owned treatment works is an issue of great importance to SOCMA's members. A large number of its members operate one or more plants which are tied to a POTW. In many cases, these plants are located in areas in which there is no reasonable alternative to the use of a POTW's services.

SOCMA's Interest

SOCMA's members are affected by the full range of regulations proposed by the EPA. However, SOCMA has a long-standing concern with the pretreatment program, and its comments focus primarily on those proposed standards.

SOCMA is particularly concerned about the regulations as applied to small organic chemical and plastic and synthetic fiber plants. Most of SOCMA's members, whether large or small companies, operate small plants. Many of SOCMA's members are small companies; approximately 55 member companies have annual sales of less than $50 million.

Small plants in the organic chemicals, plastics and synthetic fibers industries tend for a number of reasons to be indirect rather than direct dischargers. They are typically located in urban areas where land for onsite wastewater treatment systems designed for direct discharge is generally either not available or very expensive. In addition, such areas as a rule have POTWs which will accept and treat industrial wastewater. Joint treatment of domestic and industrial wastes is cost-effective. The user charges and other direct and indirect costs associated with discharging wastewater to a POTW are generally less than the cost of an onsite treatment system designed for direct discharge, particularly for small plants which cannot take advantage of economies of scale in treatment technologies.

As a result of these factors, a high proportion of small plants in the organic chemicals, plastics and synthetic fibers industries are indirect dischargers.

SOCMA's Concerns with the Present Pretreatment Strategy

SOCMA has expressed its concern that the present pretreatment strategy, founded as it is on national categorical pretreatment standards, had not led toward measurable advancement of the goals of the Clean Water Act. Since last year, EPA has made progress in developing national categorical pretreatment standards. However, many standards, including those for the organic chemical and plastic and synthetic materials industry, have not been completed. And substantial issues remain due to the diverse and complex nature of industrial users of POTWs, to the difficult technical questions relating to the treatability of particular wastewaters by POTWs and by alternative pretreatment technologies, to the difficulty in resolving important questions of POTW sludge use and disposal, and to the economic impact of those regulations, particularly on smaller companies. Further, the provisions of the general pretreatment regulations for application of the categorical standards to particular POTWs and industrial users are complicated to the point that they are, in SOCMA's view, unlikely to be a workable tool for providing needed flexibility in the national standards.

Perhaps the most important point made by SOCMA was that many municipalities already have developed effective local pretreatment programs. SOCMA believes that such programs have substantially implemented the objectives of the Clean Water Act in that they protect POTW operations, ensure compliance with POTW NPDES permit limitations, ensure safe sludge disposal, and allow cost-effective treatment of both industrial and municipal waste.

Alternative Local Pretreatment Systems as an Option to National Standards

SOCMA continues to believe that the present pretreatment strategy, with its reliance on national categorical pretreatment standards, will not be the most effective approach in advancing the goals of the Act. However, SOCMA recognizes that others have strongly objected to the total elimination of the national categorical pretreatment standards and the pretreatment strategy based on these standards.

Given the diverse views presented during last year's debate, SOCMA accepts the fact that the discussion must shift from continuation of the national standards to the development of an option to the application of these standards to industrial users of individual POTWs. The proposal for "alternative local pretreatment systems" represents a constructive effort in that direction.

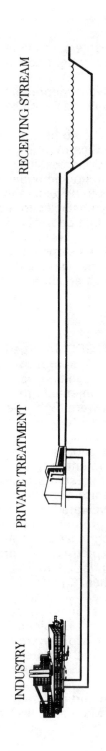

RECEIVING STREAM

PRIVATE TREATMENT

INDUSTRY

Many industries treat their own wastes in private treatment plants.

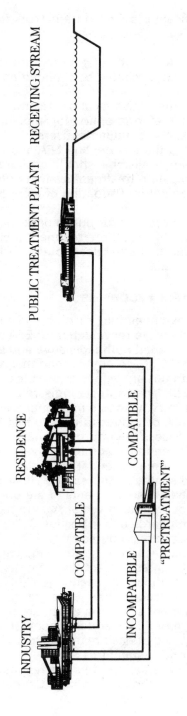

RECEIVING STREAM

PUBLIC TREATMENT PLANT

RESIDENCE

COMPATIBLE

COMPATIBLE

"PRETREATMENT"

INCOMPATIBLE

INDUSTRY

Other industries send wastes to publicly owned treatment plants.
Source: Water Pollution Control Federation

83

The concept provides a framework from which an environmentally effective and workable pretreatment strategy might be developed.

SOCMA believes that a program for alternative local pretreatment systems should be founded on the following basic principles:

First, alternative local pretreatment systems should be considered the preferred option for implementing the Clean Water Act with respect to industrial users of POTWs. The Congress, in amending the statute, and EPA and the states, in implementing such new provisions, should encourage POTWs to develop alternative local pretreatment systems. Unnecessary obstacles to the development and adoption of local systems should be avoided. . . .

An opportunity for public participation in the development of an alternative local pretreatment system should be given and, in particular, industrial users of the POTW should be encouraged to participate. . . .

SOCMA's Proposals

In its November 1981 recommendations to EPA, SOCMA urged a comprehensive reevaluation of EPA's pretreatment approach which included both legislative and administrative initiatives. However, SOCMA emphasized that, while amendments to the pretreatment provisions of the Clean Water Act were highly desirable, EPA had sufficient discretion under the Act to make significant improvements in its pretreatment program.

SOCMA continues to believe that the Agency should thoroughly assess the improvements in the pretreatment program which it can make administratively. Further, the present proposal for pretreatment standards for the organic chemicals and plastics and synthetic fibers industries should be evaluated.

The two basic elements of the pretreatment program contemplated by the Clean Water Act are national categorical pretreatment standards under Section 307(b) and local pretreatment programs developed pursuant to POTW permits under Section 402(b)(8). Each has its role in the legislative scheme.

The Congress' intent was that the "pretreatment standards established under this section [307(b)] would be national in scope and addressed to the most significant pretreatment problems." Pretreatment problems which are not national in scope were to be addressed by standards adopted by POTWs, and Congress "expected that each manager of a treatment works would provide for such standards." . . .

EPA's evaluation of its pretreatment program and of these regulations should recognize Congress' intent. More specifically,

84

EPA's pretreatment program should be based on the following:

• Development of national categorical pretreatment standards only for pollutants and for industry segments which represent a significant national pretreatment problem. . . .

• Recognition of the appropriate and necessary role of local POTWs in addressing problems which are not of national significance by support and implementation of legislation providing for alternative local pretreatment programs.

SOCMA does not believe that the regulations as proposed adequately reflect these principles. Instead, the proposed pretreatment standards reflect an apparent continuing adherence to a philosophy that national categorical pretreatment standards are the dominant, if not the only, mechanism for regulating industrial users of POTWs. SOCMA firmly believes that EPA's undue reliance on this one element of the pretreatment program envisioned by the Congress is a major cause of an Agency proposal which—

• Would impose pretreatment standards in the absence of POTW treatment data.

• Would impose pretreatment standards on the industry as a whole without a sufficient data base to identify industry segments which may pose a problem of national significance.

• Would impose pretreatment standards without identification and evaluation of a technology to meet these standards other than biological treatment which would duplicate POTW treatment.

• Would impose pretreatment standards without an adequate data base to evaluate their environmental benefit or economic impact.

• Fails to recognize the ability and responsibility of EPA, the states, and local POTWs to identify and remedy through local pretreatment programs environmental problems resulting from industrial users of POTWs which are not of national significance or capable of being remedied by categorical pretreatment standards.

ECONOMIC CONSIDERATIONS AND CLEAN WATER

Edison Electric Institute

The Edison Electric Institute is a nonprofit trade association of 189 gas and electric companies. The following statement was made by attorney John Gibson of the Pacific Gas and Electric Company on behalf of the Edison Electric Institute.

Points to Consider

1. How were water quality standards originally developed?
2. Why were some water quality criteria more stringent than needed?
3. What economic considerations are important in relation to water quality standards?
4. How are inflexible rules defined?

Reprinted from a position paper by the Edison Electric Institute, April 7, 1983.

But there may also be cases in which use designations should be revised in order to allow for economic growth or expansion.

Many states, when they first put forth their water quality standards, designated waters "fishable, swimmable" without much regard for whether such uses really made sense for particular waterways. Former EPA Administrator Douglas M. Costle pointed out this problem:

[Water quality standards] were initially developed with insufficient information regarding the practical implication of achievement; high goals of water use were set, but environmental, technological and economic constraints were seldom considered; and site-specific analysis of waterway conditions that would affect either the attainability of the use or the criteria necessary to support attainable uses was rarely conducted.

Moreover, recent studies done by the utility industry show that many of the water quality criteria established to support designated uses are also more stringent than needed. A study conducted for the Utility Water Act Group (UWAG) by Professors Arthur Buikema, Jr. and Donald Cherry at Virginia Polytechnic Institute strongly support this conclusion. Buikema and Cherry, "Evaluation of 1980 U.S. EPA Water Quality Criteria for Selected Trace Metals" (Oct. 1982). Buikema and Cherry examined EPA's 1980 criteria for several metals by comparing them to ambient concentrations reported at United States Geological Survey (USGS) benchmark stations (in streams not generally influenced by man and thus presumed to reflect natural water quality conditions) and to concentrations reported in a number of site-specific biological studies. Buikema and Cherry found that the national criteria were lower especially in soft water systems (*i.e.,* systems with low levels of calcium and magnesium). They concluded that the criteria for four of the five metals examined (cadmium, copper, lead and zinc, but not nickel) were overly stringent and could be modified with no harm to the environment.

States should not be bound by past mistakes if the passage of time, the gathering of additional data, and a careful reconsideration of water quality standards have convinced state officials that those standards need to be revised.

Economic Concerns

The clearest examples of cases in which changes to designated uses should be allowed are where natural background pollutant levels or physical limitations in the water body make the uses

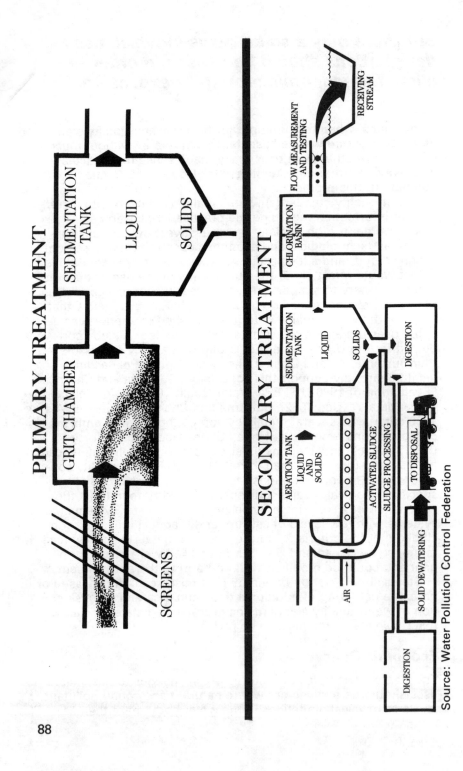

PRIMARY TREATMENT

SCREENS

GRIT CHAMBER

SEDIMENTATION TANK

LIQUID

SOLIDS

SECONDARY TREATMENT

AERATION TANK
LIQUID AND SOLIDS

AIR

ACTIVATED SLUDGE

SLUDGE PROCESSING

SOLID DEWATERING

DIGESTION

SEDIMENTATION TANK

LIQUID

SOLIDS

DIGESTION

TO DISPOSAL

CHLORINATION BASIN

FLOW MEASUREMENT AND TESTING

RECEIVING STREAM

Source: Water Pollution Control Federation

88

unattainable. But there may also be cases in which use designations should be revised in order to allow for economic growth or expansion. The need to allow for economic growth or expansion is particularly important in the case of the electric utility industry.

Steam electric power plants furnish an essential public service—indeed, electric utilities are required in most cases under state law to build the facilities necessary to provide adequate and reliable electric service for the public. The sites at which steam electric power plants can be located have traditionally been subject to a complex web of economic and social constraints stemming from such factors as proximity to load centers, to main transmission lines, and to sources of fuel, and access to methods of fuel transport. They are now also subject to important locational constraints under air pollution, water pollution and solid waste disposal laws. . . .

Inflexible Rules

The difficulties presented by an inflexible rule are multiplied if that rule is intended to prevent redesignation *within* a basic use designation (as between a cold water or warm water fishery use within the basic fish and wildlife protection use). For example, maintaining a cold water (trout) fishery designation at a particular location may require construction of cooling towers. The costs of cooling towers, which can run as high as $75 million for a single 1,000 MW plant,[1] may outweigh the benefits of maintaining the cold water fishery use. In such a case, it may make sense to change the use to a warm water (bass) fishery of equivalent social value. It is not clear whether or not this type of change in the level of aquatic protection would be prohibited, but it clearly should not be.

For all of these reasons, any rigid, arbitrary rule is unwise public policy and should not be adopted. A state should be free to designate (or redesignate) any use that is justified by the facts then before it, and by the value judgments as to competing uses that a state is best suited to make.

---

[1]Cooling towers will also have other significant adverse impacts. For example, cooling towers evaporate large quantities of water to dissipate heat to the atmosphere, consuming up to 250 percent more water than open-cycle cooling. Other adverse side effects may include fogging, icing of nearby exposed surfaces in winter, cloud formation, abnormal precipitation, salt drift, noise, aesthetic impact, and potential aviation hazards. All of these effects should be considered in a benefit-cost comparison.

Costs and Benefits

Many states set a broad goal of "fishable and swimmable," for virtually every water body in their jurisdictions. In some cases, those goals were adopted with little actual data and have proven not feasible or too costly to attain.

Kenneth E. Blower, Standard Oil Company of Ohio, April, 1983.

EPA's Proposed Water Quality Standards Regulations

EPA's recently proposed water quality standards regulations and draft guidance documents provide a realistic approach to national water quality planning. EPA's proposals represent an ambitious undertaking to provide states with comprehensive procedures for setting, revising, and applying water quality standards. On the whole, we find those proposals to be well-conceived and grounded on sound legal and policy considerations.

We especially support EPA's efforts to give states greater flexibility to revise and fine-tune their water quality standards in appropriate circumstances. It is important to recognize that EPA's proposals would not give states carte blanche to change their water quality standards, but rather would establish certain limited conditions under which states could do so. Under the proposed rules, states are encouraged to conduct analyses to determine whether designated uses are attainable. If they are not attainable because of physical limitations or natural background levels of pollutants, states are allowed to change those uses. States may also change designated uses if they determine that the benefits of attaining the uses do not bear a reasonable relationship to the costs.

By giving states greater flexibility to establish and revise designated uses, EPA's proposals advance policy to preserve and protect the primary responsibilities and rights of states in pollution control. This policy is especially important with respect to water quality standards, since designating the use of a particular water body is a complicated matter of social and political judgment calling for the weighing of competing goals and the allocation of limited resources among competing interests, much like land-use planning and zoning. States are in the best position to make these judgments since they are most familiar with local conditions and values.

WHAT IS POLITICAL BIAS?

This activity may be used as an individualized study guide for students in libraries and resource centers or as a discussion catalyst in small group and classroom discussions.

Many readers are unaware that written material usually expresses an opinion or bias. The skill to read with insight and understanding requires the ability to detect different kinds of bias. Political bias, race bias, sex bias, ethnocentric bias and religious bias are five basic kinds of opinions expressed in editorials and literature that attempt to persuade. This activity will focus on political bias defined in the glossary below.

5 KINDS OF EDITORIAL OPINION OR BIAS

*sex bias—the expression of dislike for and/or feeling of superiority over a person because of gender or sexual preference

*race bias—the expression of dislike for and/or feeling of superiority over a racial group

*ethnocentric bias—the expression of a belief that one's own group, race, religion, culture or nation is superior. Ethnocentric persons judge others by their own standards and values

*political bias—the expression of opinions and attitudes about government-related issues on the local, state, national or international level

*religious bias—the expression of a religious belief or attitude

Guidelines

Read through the following statements and decide which ones represent political opinion or bias. Evaluate each statement by using the method indicated below.

Mark (P) for statements that reflect any political opinion or bias.

Mark (F) for any factual statements.

Mark (O) for statements of opinion that reflect other kinds of opinion or bias.

Mark (N) for any statements that you are not sure about.

91

___ 1. The Clean Water Act of 1972 gave the federal government primary responsibility for enforcing water pollution controls.

___ 2. Water pollution controls should be the primary responsibility of state governments.

___ 3. The Clean Water Act has led to the clean up of some of our most polluted waters in the nation.

___ 4. It is more important to have a healthy economy than a clean environment.

___ 5. Clean water should take precedence over protecting industries and jobs that pollute our lakes and streams.

___ 6. Maintaining clean lakes and rivers promotes a higher standard of living.

___ 7. A majority of Americans favor strong restrictions on water polluting industries no matter what the economic costs or consequences.

___ 8. Most controversies surrounding the issue of clean water involve who will pay the cost of technologies and equipment to stop the pollution.

___ 9. Polluting industries should pay for safe and non-polluting methods of toxic waste disposal.

___10. About 29% of the groundwater drinking water supplies of 954 U.S. cities with populations over 10,000 are contaminated.

___11. The vast majority of the estimated 700 chemicals so far detected in drinking water have not yet been tested to determine toxicity.

___12. Today nearly 12 million Americans do not have access to clean water.

___13. The World Health Organization finds that 8 of 10 cases of fatal illnesses are caused by contaminated water supplies.

___14. The Ogallala aquifer, the largest in the world, is being drained to grow crops the nation does not need.

___15. It is appropriate to place industrial wastes in groundwater that will not be needed for drinking water.

Other Activities

1. Locate three examples of political opinion or bias in the readings from chapter two.
2. Make up one statement that would be an example of each of the following: **sex bias, race bias, ethnocentric bias,** and **religious bias.**

CHAPTER 3

GROUNDWATER CONTAMINATION

READINGS

GROUNDWATER CONTAMINATION

MAJOR SOURCES OF GROUNDWATER CONTAMINATION

Donald V. Feliciano

The following statement presents an overview of groundwater contamination and the major legal and political issues surrounding attempts to regulate pollution on the state and federal levels. Donald V. Feliciano is a staff researcher for the environment and natural resource division of the Library of Congress. For a description of the 33 major sources of groundwater contamination, see appendix.

Points to Consider

1. What are the three major sources of groundwater contamination?
2. How are the major groundwater issues facing Congress described and identified?
3. How large are the U.S. groundwater reserves?
4. What are the best ways to protect groundwater?

Reprinted from a pamphlet titled *Groundwater Contamination,* Library of Congress, January, 1985.

Groundwater reservoirs, or aquifers, contain nearly 50 times the volume of the Nation's surface waters, constitute 96% of all the fresh water in the United States, and are the primary drinking water source for half of the population—nearly 115 million people.

Groundwater contains over 50 times the volume of the Nation's surface waters and provides over half the U.S. population with its primary source of drinking water. Increasing concern is being expressed because groundwater in some locations is contaminated by toxic or potentially hazardous chemicals, many of which are known or suspected carcinogens.

No single piece of legislation serves to protect this resource exclusively. The U.S. Environmental Protection Agency (EPA) protects the drinking water quality of groundwater through several evironmental laws, but they overlap in parts, leading to some duplication of effort. Furthermore, they are not comprehensive, even when administered to complement individual State protective efforts. EPA has attempted to integrate the administration of the Federal and State laws by issuing a comprehensive national groundwater protection strategy, but this strategy is in the early stages of implementation.

Groundwater issues facing Congress include whether groundwater contamination is being monitored and evaluated sufficiently and whether Federal involvement in groundwater protection should be strengthened. . . .

Background and Policy Analysis

Groundwater is one of our largest natural resources and probably the least understood by the public. Located underground and usually within 2,500 feet of the surface, groundwater reservoirs, or aquifers, contain nearly 50 times the volume of the Nation's surface waters, constitute 96% of all the fresh water in the United States, and are the primary drinking water source for half of the population—nearly 115 million people.

Sources of Groundwater Contamination

Every State in the United States has experienced some level of groundwater contamination, the sources of which generally fall into three categories.

(1) Natural pollutants—mineralization and saltwater encroachment of aquifers. In arid areas mineralization may result via

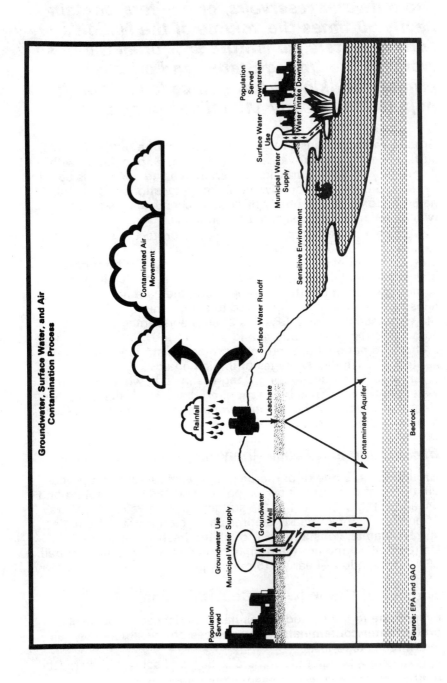

Groundwater, Surface Water, and Air
Contamination Process

Population Served

Groundwater Use
Municipal Water Supply

Groundwater Well

Rainfall

Contaminated Air Movement

Surface Water Runoff

Leachate

Contaminated Aquifer

Sensitive Environment

Surface Water Use
Municipal Water Supply

Population Served Downstream

Water Intake Downstream

Bedrock

Source: EPA and GAO

96

leaching from the substrate through which the water passes. Evaporation further concentrates these mineral salts. In the southwest and southcentral areas of the country, natural contamination of groundwater is the primary problem affecting the resource. In the coastal areas, and particularly where aquifers have been overdrafted, saltwater from the ocean also may invade and contaminate the groundwater.

(2) Human waste disposal activities—industrial waste impoundments and landfills, municipal landfill operations, underground injection operations, and individual septic tank systems. As of 1982, EPA located over 180,000 waste impoundments at some 80,000 sites. Of the industrial sites evaluated in the agency's 1980 Surface Impoundment Assessment, 70% were found to be unlined, 50% were sitting directly on top of aquifers used as sources of drinking water, and 98% were located within one mile of a water supply well. EPA believes that there are 75,000 operating industrial landfills containing nonhazardous waste and about 200 hazardous waste landfills.

As of 1984, EPA estimates that there are over 18,000 uncontrolled hazardous waste sites in the United States and that nearly 2,200 may require Federal cleanup attention to protect public health.

Municipal landfills also indicate potentially hazardous conditions. According to the 1983 Survey of Landfills conducted by *Waste Age Magazine,* of 12,991 operating municipal landfills identified in the survey, 2,396 were open dumps, only 37 had artificial liners, and only 1,609 had monitoring wells to detect groundwater contamination. There are also some 24,000–36,000 inoperative municipal landfills.

There are over 261,000 underground injection wells in this country into which a wide range of chemical wastes are being disposed. The practice is based on the assumption that wastes can be safely injected into deep, confining geologic strata or saline aquifers with no intended uses, but existing knowledge of the implications of underground injection is scanty. Groundwater contamination has occurred near injection wells, and shifting within the geologic strata has the potential of freeing confined wastes into the surrounding groundwater.

Regarding private septic systems, there are an estimated 19.5 million homes in the U.S. using such facilities, and their total estimated discharge of liquid wastes into the ground exceeds 1 trillion gallons per year. Until recently, a popular way to clean out septic systems instead of pumping was to use a solvent containing trichloroethylene. Also known as TCE, it is one of the most common pollutants of groundwater.

(3) Human activities other than waste disposal—runoff from agriculture, mining, oil and gas, and silviculture operations,

Major Sources of Groundwater Pollution

Major sources of ground water pollution include:
- 22,000 abandoned hazardous waste disposal sites
- 31,000 unregulated industrial pits, ponds and lagoons containing hazardous wastes;
- 250 to 300 hazardous waste injection wells;
- 1.4 to 2.0 million underground storage tanks;
- 15,000 active municipal landfills;
- 1,500 hazardous waste landfills;
- farm and home use of pesticides and fertilizers;
- animal feedlots;
- highway de-icing salts;
- septic tanks.

Not Man Apart, October, 1984.

roadway de-icing agents, acid precipitation, and accidental spills, including leakage from underground storage tanks. Many of the contaminants constitute what are known as non-point sources (that is, not coming from a pipe) of pollution. Because their point of entry into an aquifer is very difficult to determine, methods to control this pollution are complex. Regarding accidental spills, several experts contend that some of the largest contributors to groundwater contamination are leaking underground pipelines and storage tanks, such as gasoline tanks underneath automobile service stations. Underground storage tank estimates range from 500,000–1 million (nonpetroleum tanks) to 2.5 million (petroleum tanks). . . .

Extent of Groundwater Contamination

Although it is not yet possible to assess groundwater contamination on a national level, evidence of the extent of the problem is accumulating. The Council on Environmental Quality estimated in 1981 that serious contamination of groundwater by toxic synthetic organic chemicals had occurred in 34 States. Within the last few years, for example, public wells have been closed in 22 communities in Massachusetts, 29 in Connecticut,

25 in Pennsylvania, and 60 in New York. Similarly, over 100 private wells have been closed in New Jersey and 500 in Long Island. In the San Gabriel Valley in California, 107 of 349 public wells serving 4.5 million people have significant levels of trichloroethylene, an organic chemical known to cause cancer in mice. Also, a 1983 CRS summary of contaminated drinking water wells, taken from State reports, indicates that 2,820 wells have been either affected or closed as a result of toxic substance contamination in the last few years. Finally, there is the threat to groundwater from abandoned hazardous waste sites: of the 786 sites now listed or proposed on the Superfund priority list, 585 (74%) appear to have groundwater contamination. . . .

Experts believe that the best way to protect groundwater is to: (1) eliminate the contaminants at their source; (2) protect the recharge areas of unconfined aquifers by regional planning; and (3) develop an effective monitoring system so that contamination of groundwater is quickly spotted.

Policy Issues

(1) *What is the status of information about groundwater contamination?* Groundwater contamination is such a "new" area of environmental concern that the problem and its extent are still being evaluated. National data on groundwater contamination are limited, and there is no national monitoring system to determine the degree or trends of groundwater degradation. And, although there are numerous regional, local, and private reports on groundwater quality, there is no central collected source for this information. EPA admits that it is still in the rudimentary stages of defining the problem, and many States are just beginning to respond to concerns about groundwater. Much of this concern has been magnified by contamination threats resulting from hazardous waste sites.

Even though only an estimated 1% of the Nation's groundwater is contaminated, thousands of people already have been exposed to this contamination prior to the closing of their drinking water wells. Further research is needed to determine the human health effects from drinking water that may be contaminated, particularly in the area of epidemiology. Also, there has been little research on how to clean up contaminated aquifers.
. . .

(2) *How well are the existing Federal laws that address groundwater protection working?* The Federal Government already has expressed the need for groundwater protection as components of several environmental laws, yet recent evidence has called into question how well these laws are protecting groundwater. . . . For example, an EPA study completed in January 1983 found a 64% lack of compliance with RCRA ground-

water monitoring regulations; a follow-up study by EPA released in 1984 showed less than 20% compliance. Similarly, the experience of the Pine Barrens groundwater recharge areas of the Long Island sole source aquifer has shown that the Safe Drinking Water Act does not provide comprehensive management of sensitive aquifers.

Because groundwater is protected in part by several environmental acts, there may be a need for Congress to strengthen the Federal oversight of groundwater. In this regard, two alternatives may be considered: amend each environmental act to be more responsive to groundwater protection (perhaps by emphasizing integration of the various statutes), or enact a groundwater-specific protection law.

Although no groundwater-specific bill has been introduced in Congress, bills have been introduced in the 99th Congress to amend the Safe Drinking Water Act (S. 24, S. 124) and the Clean Water Act (H.R. 8) to be more responsive to groundwater protection. The RCRA reauthorization act (P.L. 98–616) contained several groundwater provisions. . . .

(3) *Do the Federal and State roles in groundwater protection need to be revised?* Groundwater protection has traditionally been a right of the States; nevertheless, the Federal role has been growing for several reasons. One reason is that groundwater aquifers are interstate. In some cases the recharge area of an aquifer (and sometimes the source of contamination) occurs in a different State from the one withdrawing groundwater from the aquifer.

A second reason is that some existing Federal environmental statutes intrinsically affect groundwater contamination: accelerated cleanup in one medium (air, water, or land) generally means that pollutants will be transferred to other media, and usually the least regulated. A third reason stems from the diversity of State laws protecting groundwater. Generally reflecting the groundwater uses within a particular State, they provide little integration and consistency among the State efforts.

The ultimate questions to managing groundwater that Congress faces include:

—What should be the balance of Federal-State roles in groundwater protection?

—What are the consequences of protecting groundwater through a preventive approach, as compared to reacting to groundwater contamination as is the usual case today?

EPA GROUNDWATER
STRATEGY: THE POINT

Environmental Protection Agency

The following statement is excerpted from the Environmental Protection Agency's Groundwater Protection Strategy issued in August of 1984. It has four major national goals and places the major regulatory responsibilities on state and local governments.

Points to Consider

1. Who is responsible for protecting groundwater?
2. Why are state and local officials given primary responsibility for protecting groundwater?
3. What are the three classes of groundwater protection?
4. What are the four major goals of the EPA groundwater protection strategy?

Environmental Protection Agency, *Groundwater Protection: The Quest for a National Policy,* August, 1984.

EPA Ground-Water Protection Strategy represents a major step forward in EPA and State efforts to protect ground water.

In the last decade the public has grown increasingly aware of the potential problem of ground-water contamination. Reports of chemicals threatening drinking water supplies have mobilized State, local and Federal governments to respond. But these responses suffer from a lack of coordination among responsible agencies, limited information about the health effects of exposure to some contaminants, and a limited scientific foundation on which to base policy decisions. Officials at all levels of government have begun to look for a definable strategy to protect ground water. The strategy presented here will provide a common reference for responsible institutions as they work toward the shared goal of preserving, for current and future generations, clean ground water for drinking and other uses, while protecting the public health of citizens who may be exposed to the effects of past contamination. . . .

Who is Responsible for Protecting Ground Water?

The Ground-Water Protection Strategy was developed in full recognition of EPA's recently released policy statements on delegation and oversight. The clear intent of those policies is to make use of Federal, State, and local governments in a partnership to protect public health and the environment. State and local governments are expected to assume primary responsibility for the implementation of environmental programs because they are best placed to address specific problems as they arise on a day-to-day basis. The EPA role is to provide national environmental leadership, develop general program frameworks, establish standards required by Federal legislation, conduct research and national information collection, provide technical support to States, and provide assistance to States in strengthening their programs. The Federal, State, and local roles as expressed in this Strategy are completely consistent with EPA's delegation and oversight policies.

The EPA role identified above will involve cooperation from other Federal agencies, especially regarding research information collection and technical support to the States. The EPA will provide program leadership and technical assistance to the States in matters involving the protection of ground-water quality, and will rely on the Department of the Interior and the U.S.

102

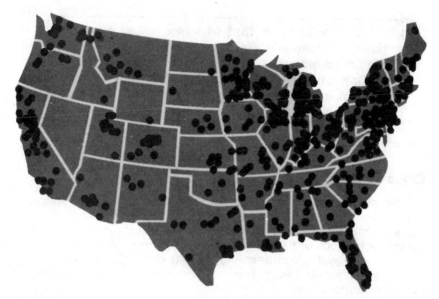

Source: EPA
Areas Containing Major Hazardous Wastes Sites

Geological Survey (USGS), for assistance in defining the hydraulics and geochemistry of ground-water flow.

Guidelines for Ground-Water Protection

EPA's guidelines for ground-water protection are based on the beneficial use criterion. Protection policies are defined for three classes of ground water. The class definitions reflect the value of the ground water and its vulnerability to contamination, and they apply to ground water having significant water resources value. The three classes are: I) Special Ground Water, II) Current and Potential Sources of Drinking Water and Water Having Other Beneficial Uses; and III) Ground Water Not a Potential Source of Drinking Water and Having Limited Beneficial Use. These guidelines establish the basic framework for the Ground-Water Protection Strategy.

In describing the various classes of ground water, emphasis is on broad definitions and basic criteria to be used in class assignment. Guidance will be developed establishing more specific criteria and definitions for classifying ground water. This guidance may prescribe additional criteria to be used in identifying the various classes.

> **EPA recognizes that in some cases alternatives to ground water cleanup and restoration may be appropriate. In these cases the contamination may be managed in order to avoid migration into a current source of drinking water or to avoid widespread damage. . . .**

Class I—Special Ground Waters

Certain ground-water resources are in need of special protective measures. These resources are defined to include those that are highly vulnerable to contamination because of the hydrogeological characteristics of the areas under which they occur. Examples of hydrogeologic characteristics that cause ground water to be vulnerable to contamination are high hydraulic conductivity (Karst formations, sand and gravel aquifers) or recharge conditions (high water table overlain by thin and highly permeable soils). In addition, special ground waters are characterized by one of the following two factors:

(1) Irreplaceable source of drinking water. These include ground water located in areas where there is no practical alternative source of drinking water (islands, peninsulas, isolated aquifers over bed rock) or an insufficient alternative source for a substantial population; or

(2) Ecologically vital, in that the ground water contributes to maintaining either the base flow or water level for a particularly sensitive ecological system that, if polluted, would destroy a unique habitat (e.g., those associated with wetlands that are habitats for unique species of flora and fauna or endangered species).

In order to prevent contamination of special ground waters, EPA will use RCRA authorities to initially discourage by guidance, and will eventually propose regulations to ban the siting of new hazardous waste land disposal facilities above these ground waters. . . .

Class II—Current and Potential Sources of Drinking Water and Water Having Other Beneficial Uses

All other ground water currently used or potentially available for drinking water and other beneficial use is included in this category, whether or not it is particularly vulnerable to contamination. This comprises the majority of usable ground water in the United States.

As a general rule, Class II aquifers will receive levels of protection consistent with those now provided for ground water under EPA's existing statutes. This means that prevention of contamination will generally be provided through application of design and operating requirements based on technology, rather than through restrictions on siting, though exceptions may apply. Cleanup of contamination will usually be to background levels or drinking water standards, but alternative procedures may be applied for potential sources of drinking water or water used for agricultural or industrial purposes. EPA recognizes that in some cases alternatives to ground water cleanup and restoration may be appropriate. In these cases the contamination may be managed in order to avoid migration into a current source of drinking water or to avoid widespread damage. . . .

Class III—Ground Water Not a Potential Source of Drinking Water and of Limited Beneficial Use

Ground waters that are saline or otherwise contaminated beyond levels which would allow use for drinking or other beneficial purposes are in this class. . . .

Prevention of contamination may be less than that provided for Class I or II in some instances, but high levels of protection will still be required in other cases. New and existing hazardous waste land disposal facilities regulated under RCRA will be required to meet the same technical standards—such as liners, leachate collection systems, and monitoring—as facilities located over Class I or II ground water. Hence, in terms of protection, the Ground-Water Protection Strategy currently envisions the same technical standard of protection for hazardous waste facilities in all classes. With respect to cleanup, should the hazardous waste facility leak, the Agency would normally grant variances that establish elevated concentration limits. Generally, cleanup decisions for Class III ground water that has been contaminated by human activities would be evaluated on a case-by-case basis. If contamination poses no risk to human health and the environment, as will frequently be the case (because the ground water is not usuable and there are controls to ensure it is not used), then, under RCRA, cleanup requirements could be reduced or eliminated.

Summary

In summary, this EPA Ground-Water Protection Strategy represents a major step forward in EPA and State efforts to protect ground water. The strategy has four goals:
- **To foster stronger State government programs for Ground-**

Water Protection:
—EPA will provide grant support for State program development;
—EPA will offer technical assistance to States; and
—EPA will target research efforts to State requirements.

- **To cope with inadequately addressed problems of ground-water contamination:**
—EPA will assess the extent of contamination by leaking underground storage tanks, issue a Chemical Advisory warning to gasoline station owners and operators of the problem, and consider the need to further regulate these contamination sources;
—EPA will assess the problems associated with surface impoundments and landfills; and
—EPA will strengthen its efforts to protect ground water from pesticide contamination and over time assess the effects of other practices on ground-water quality.

- **To establish a framework for decision-making by EPA programs:**
—EPA will adopt guidelines for ground-water protection. These guidelines will assure a high level of protection for ground water used for drinking and other beneficial purposes, and bring about greater cohesion in EPA ground-water protection efforts.

- **To strengthen the internal ground-water organization**
—EPA has established an Office of Ground-Water Protection in the Office of Water and counterpart offices will be established in each Region.

EPA believes that this strategy represents a pragmatic, evolutionary approach to improving the protection of the Nation's ground-water resource. It will provide the institutional muscle needed to bring about the needed change. It provides at the Federal and State levels a framework for decision-making and a roadmap to address new problems. This Strategy can only be successful through EPA leadership, the development of strong State programs, and general support from Congress, environmentalists, and the regulated community.

EPA GROUNDWATER STRATEGY: THE COUNTERPOINT

Environmental Defense Fund

The Environmental Defense Fund is a national not-for-profit, public interest environmental organization with over 46,000 members. These comments address the Environmental Protection Agency's Ground-Water Protection Strategy of 1984.

Points to Consider

1. What is the major weakness of the EPA's groundwater protection strategy?
2. How should the role of the states be defined?
3. What were the unaddressed threats to groundwater?
4. What is wrong with the aquifer classification scheme?

Excerpted from a position paper issued by the Environmental Defense Fund, March 30, 1984.

We fail to share EPA's confidence that many states will be willing or able to divert a significant portion of grant funds now dedicated to other programs to the effort.

EPA's draft Ground-Water Protection Strategy properly recognizes that protection of the nation's ground-water resources from contamination should be a major environmental priority today. . . . It also recognizes that enormous resources will be required to prevent future contamination and to detect and address existing ground-water problems.

Unfortunately, the strategy does not outline a significantly increased commitment of resources to ground-water protection. For the most part, the strategy focuses on improving coordination of existing programs while gathering further information about the nature and extent of the problem and possible regulatory responses. Both are important activities and EPA's recognition that the ground-water problem is serious enough to require a coordinated response is particularly encouraging. But in the absence of a significant increase in the resources committed to ground-water protection, the threat to human health and the environment from ground-water contamination will continue to increase. We urge EPA to "put its money where its mouth is" and to directly address the issues of how much resources are needed to do the job and where these resources will come from. . . .

Role of States

Financial support for program development is absolutely essential, but we fail to share EPA's confidence that many states will be willing or able to divert a significant portion of grant funds now dedicated to other programs to the effort. Implementation funds will also be necessary at some point and the Agency must realize and begin to act on the fact that existing grant programs will be inadequate to provide effective assistance at the implementation stage. In sum, while we agree with the Agency's move to enhance state capabilities, we are not confident that the level of effort and assistance proposed will be adequate to attain and continue the necessary program capability at the state level.

The questionable effectiveness of the rather weak "stick and carrot" approach. While EPA proposes to use grant funds that are available under existing programs to assist states in pro-

HEAD in The SAND

Reprinted with permission of *The Minneapolis Star and Tribune*

gram development, it does not explain how it will persuade states to agree to the diversion of those federal funds from their other programs. It does not discuss the levels of funding that will be necessary to induce states to invest their own resources in a new program, and it does not attempt to examine the enormous difficulty inherent in inducing the states to develop a program that they alone will apparently have to fund beyond the development state. Despite the assistance offered by EPA, the states will be left to shoulder a tremendous financial burden in order to administer what will be a complex and difficult program. Furthermore, a number of state governments will have to overcome considerable political hurdles in order to impose new regulatory requirements, hire new personnel, and expand their budgets.

Of course, these concerns are present in all of the federal environmental programs enacted over the past 15 years. What makes the difference is that in many of those programs EPA has a credible stick to go along with the funding carrot. That is, a state's unwillingness or inability to administer the program means federal administration. With respect to the groundwater strategy, EPA makes some half-hearted, ceremonial noises about using its leverage over what it refers to as "delegated programs" to induce state compliance. Perhaps with respect to those states that are still awaiting authorization, EPA will have

109

some leverage and can require adoption of certain elements of the strategy, such as the hazardous waste site locational standards, as a prior condition. If that is the intent, the speedy development of the federal guidelines obviously becomes crucial. EPA will not be able to hold up pending applications (e.g., for final authorization under RCRA) in order to develop the groundwater classification system.

It is more likely to assume that most of the programs will already have been authorized, and that inclusion of the ground water strategy elements will require states to amend their programs. Presumably, failure to conform the program might be considered justification for EPA to rescind authorization, but we doubt that this is considered to be a credible deterrent any more. The Agency does not assert very convincingly that it will withhold authorization, much less rescind it, for failure to implement the groundwater strategy.

Neither the carrot nor the stick proposed by EPA is sufficient to this task. EPA must seriously consider at least obtaining adequate financial assistance to make the program work. And, it must make the case for groundwater protection strongly enough to overcome political resistance at the state level to an expanded regulatory program. . . .

Unaddressed Threats to Ground-Water

While we are encouraged that EPA plans to address unaddressed sources of groundwater contamination in its strategy, we are concerned that the scope of the program presented is too narrow. Our specific criticisms fall into two categories:

1. *The action proposed for the two currently suspected significant sources of contamination—underground storage tanks and industrial surface impoundments—is not adequately aggressive.* For example, EPA's plans to regulate underground storage tanks center largely around a study to identify the nature, extent, and severity of groundwater contamination from leaking tanks. Although we support the Agency's desire to develop a data base for any regulations it issues, we believe that adequate information already exists to determine the nature and scope of this problem. This information has been generated by the states, many of whom are consequently moving on a vastly accelerated pace to address this problem. . . .

The last 5 to 10 years have witnessed major advances in the technology and management practices available for the storage and handling of hazardous liquids. New tank designs and tank materials have been developed which solve the problems of corrosion and prevent leaks. Mechanical and electronic flow control and level detection devices have been invented to prevent transfer spills. Secondary containment designs have been developed.

Again, many of these technologies have been well studied by states. (See, for example, Technology for the *Storage of Hazardous Liquids,* New York State Department of Environmental Conservation, Albany , NY, January 1983.) Thus, we urge EPA to de-emphasize the study phase and accelerate regulatory development. Clearly, the information needed to do this is already at hand. . . .

2. *Other potential sources of groundwater contamination should be given high priority for study.* Numerous potential sources of contamination to groundwater exist which have never been studied to the extent of tanks and impoundments. Pesticide and herbicide application, for example, is currently largely overlooked as a contamination source, as are many other nonpoint sources.

A most useful function that a national groundwater strategy could serve would be to systematically identify these currently unquantified potential sources of groundwater contamination and then to undertake a source sampling program to begin to estimate their national impact. EPA's current overlooking of these less obvious sources will be inefficient in the long-run, as they may lead to large future problems which will have to be addressed by costly corrective action.

EPA's Guidelines for Ground-Water Protection

Use of Guidelines—The draft strategy outlines a set of "Guidelines for Ground-Water Protection" which are designed to help ensure consistency in decisions made by various EPA Programs. The guidelines emphasize that the value of the ground-water resource to be protected should be taken into account in decisions concerning ground-water protection and cleanup priorities. Thus they specify different levels of protection and cleanup for three different categories of aquifers.

While the draft strategy states that the guidelines are designed to ensure consistency among EPA programs, consistency does not require different levels of protection for different categories of aquifers. The strategy recognizes as much when it states that the Agency considered whether all groundwater should be protected equally. Thus, the usefulness of the aquifer classification scheme employed in the guidelines must be assessed on its own merits.

Aquifer Classification Scheme—The general principle that environmental protection resources should be allocated where they wil do the most good is entirely unobjectionable. But, it does not imply that standards of protection should vary significantly based on the current uses of aquifers. The classification scheme adopted in the guidelines is of concern because it provides special protection only to an extremely narrow class of aquifers while authorizing reduced standards of protection for aquifers not considered potential sources of drinking water.

The definition of special aquifers is far too narrow. It would provide special protection only to aquifers especially vulnerable to contamination in circumstances where such contamination could either damage irreplaceable sources of drinking water or destroy unique habitats. Significant sources of drinking water and valuable habitats that are vulnerable to contamination should be afforded special protection even if they can be replaced or are not totally unique.

For the second category of aquifers (current and potential sources of drinking water) the guidelines state that cleanup standards will vary depending primarily upon whether the groundwater is used currently for drinking water. There is no reason, however, for not applying the same standard of "highest technically feasible level required to protect human health and the environment" both to current and potential sources. EPA does not have the ability to predict accurately future uses of aquifers and should therefore be concerned with all aquifers of drinking water quality. We therefore strongly object to any scheme which differentiates between current and future uses.

There are significant dangers in authorizing reduced levels of protection for aquifers not considered potential sources of drinking water. Given the tremendous uncertainties that surround the migration of contaminants, it will be extremely difficult, if not impossible, to ensure that contaminants do not migrate to sur-

112

face water or Class I or II groundwaters. For this reason, we urge EPA to distinguish aquifers largely for the purposes of levels of cleanup rather than providing for differential levels of protection. For example, facility design should always be state-of-the-art, rather than something short of state-of-the-art for Class III aquifers, as currently proposed. Additional protection via prohibitions of certain types of facilities could be considered above and beyond the uniform design requirements for Class I aquifers.

GROUNDWATER MANAGEMENT: INDUSTRIAL PERSPECTIVES

Steven I. Werner

Steven I. Werner is currently Operations Manager for Weston Designers and Consultants in West Chester, Pennsylvania. Ground water concerns were an important part of his job in the Corporate Environment Health and Safety group at Occidental Petroleum Corporation. In addition, he was the past chairman of the Chemical Manufacturers Association's Ground Water Task Group. In this capacity, he worked actively with many people in the chemical industry, other trade associations and the Environmental Protection Agency in developing concepts for the protection of this valuable natural resource.

Points to Consider

1. What should be the relationship between the federal government and the states in groundwater protection?
2. How is the issue of water quality vs. water quantity explained?
3. What should be the major goal of a groundwater protection plan?
4. What have chemical companies done to protect groundwater?

Reprinted from a statement requested by the publisher, March 3, 1985.

Individual states should have the primary responsibility for implementing ground water policy.

Management of ground water is as old as civilization itself. Almost five thousand years ago people captured and controlled this resource for irrigation and drinking water without any knowledge of hydrogeology. To supplement scarce water supplies, ancients constructed dams, developed underground channels for transporting water, and constructed wells. Utilization of ground water greatly preceded understanding of its origin, occurrence, and movement. . . .

Today, we have a great body of knowledge about ground water, enhanced by mathematics, soil chemistry and mechanics, models and a data base developed from thousands of wells drilled for water supply and ground water monitoring. . . .

A National Policy

A national ground water policy requiring federal environmental laws covering ground water to be administered in a consistent fashion is needed. Congress has already taken several important steps toward protecting our ground water resources. Major environmental legislation has been enacted, designed to control potential discharges to aquifers. The Clean Water Act, Resource Conservation and Recovery Act, Safe Drinking Water Act, and the Comprehensive Environmental Response, Compensation and Liability Act (Superfund) are reducing and controlling industrial, commercial and municipal discharges that contribute to ground water contamination. These control measures, however, do not yet thoroughly address nonspecific source discharges resulting in area wide ground water problems. . . .

Many states presently have water pollution control statutes that extend to ground water. Many states also have specific statutory authority to develop ground water management systems. Most of the western states have implemented general permit systems for allocating the quantity of ground water. A few eastern states have followed suit, although quantity is generally not a priority because of an abundant water supply. Very few states have implemented or adopted ground water policies that comprehensively address both the management of quantity and quality of ground water resources within their jurisdiction.

Individual states should have the primary responsibility for implementing ground water policy. The federal role in ground water policy should be to identify a broad national goal, develop

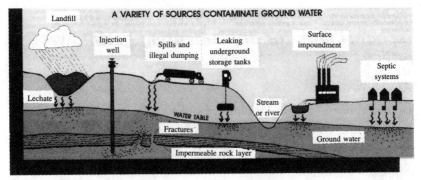

A VARIETY OF SOURCES CONTAMINATE GROUND WATER

Landfill — Injection well — Spills and illegal dumping — Leaking underground storage tanks — Surface impoundment — Septic systems — Lechate — WATER TABLE — Stream or river — Fractures — Ground water — Impermeable rock layer

Source: EPA

use classes, and provide supplemental technical and financial assistance to the states sufficient to implement their management programs. The states have the sole right of water allocation and therefore the sole right to assign the use classes and determine appropriate management techniques. The states should also be aware of the need for regional cooperation. The federal government role should be one of advisor, funder and supplier of technical assistance and scientific information.

Our nation's ground water supply is huge. By promptly enacting a comprehensive management system, we will be able to preserve that resource for all of its potential uses. The Environmental Protection Agency has been working to develop a ground water strategy and has determined that ground water will be one of its priority activities. . . .

Quality vs. Quantity

A sound ground water management policy must take into consideration the issue of water quality versus quantity. The strategy addresses only quality issues and indicates that quantity matters should be left for the states. Certainly the Western states would take issue with any Federal agency attempting to manage their ground water supplies. However, a comprehensive management system can not be achieved by focusing only on quality. A system based on water quality alone does not adequately recognize other important uses of this resource. For example, the quantity of available ground water is typically the paramount concern for such activities as agriculture, industry, mining and energy development. These uses do not always require the same quality as drinking water. Indeed, a comprehensive approach must also include other important criteria, such as yield and availability, alternate sources of water and differing uses of ground water. . . .

The goal of any National Ground Water Policy should be to protect human health and the environment and to responsibly maintain multiple uses of the resource. Hydrogeologic systems vary in accessibility, quantity and quality throughout the country. Even without human impact, certain ground waters are unsuitable for human consumption, industrial or agricultural uses because of naturally high levels of chemical or biological substances.

A ground water management policy must be developed that allows innovative and flexible techniques to accommodate the diverse nature of occurrences, uses, and existing qualities of the nation's ground water. Simplistic solutions, such as a blanket non-degradation policy, will not achieve the multiple-use objective.

There are specific instances, however, in which multiple-use is not applicable. Specifically, where there is a sole-source aquifer, that aquifer must be protected, and potential routes of contamination should be minimized. Degradation of ground water quality beyond that appropriate for the use classification should not be allowed. . . .

Several estimates indicate that less than 1% of the nation's ground water is contaminated. Most of this nation's ground water exists in its natural state varying widely in quality and quantity. Furthermore, it must be remembered that what contaminated means, depends on desired use.

The recommended management approach to achieve the stated goal is to identify use classes, and develop a comprehensive data base on ground water contaminants and sources of ground water pollution. The data base would be used to designate classes of ground water.

The management of the nation's ground water must recognize that the resource is needed to accommodate a variety of societal interests. These interests include use of water for human consumption, livestock watering, irrigation, industrial water supply, geothermal applications and waste water disposal. . . .

The individual states should have the primary responsibility for implementing ground water policy. As indicated earlier, the federal role in ground water policy should be to identify a broad national goal, develop use classes, and provide supplemental technical and financial assistance to the states sufficient to implement their management programs.

The states have the sole right to allocate waters within their jurisdiction. The states should, therefore, retain the sole authority to assign use classes to ground water and to determine the management techniques needed to maintain those assigned uses. The occurrence of ground water and its value vary. It will be important for each state to allocate and manage the resource

117

within its jurisdiction as well as to recognize the need for regional cooperation. Any federal involvement must not de facto abrogate the state's authority.

The Chemical Industry

One trade association, the Chemical Manufacturers Association (CMA) has taken a lead role in ground water issues. During the past four years it has worked diligently with other trade associations, EPA and other federal and state agencies to foster sound ground water management. Through the Ground Water Management Task Group and the recently formed ad hoc Ground Water Strategy Group, CMA has explored the issues, published papers and provided comment on proposed legislation and regulation.
. . .
A great many companies in the chemical industry have developed voluntary programs to address ground water issues at manufacturing locations. These include hydrogeologic investigations and the institution of corrective measures where required.

One company has developed the concept of a Ground Water Protection Plan. The plan proposes that each facility having unregulated potential sources of ground water contamination prepare and implement a site-specific Ground Water Protection Plan. One of the principal advantages of this approach is that the program would be self-implementing. The concept would be similar to existing Spill Prevention, Control and Countermeasure Plans under the Clean Water Act. It could also avoid long delays

118

and drains on agency resources that would be part of a mandated regulatory program.

An issue of great importance to the chemical industry is Leaking Underground Storage Tanks. Some companies are developing sophisticated techniques to detect the potential for leaks before they actually occur. One successfully employed technique involves an extrapolated form of classical acoustic emissions. This inspection methodology is a nondestructive technique capable of evaluating the integrity of a structure. The technique assesses the physical integrity of a structure by listening to acoustic emissions which are generated by regions of discontinuity in a structure under increasing stress. This information is processed and used to detect and locate flaws and prevent potential failures.

A recent survey conducted by CMA has confirmed a move away from reliance on landfills and increased use of waste destruction methodologies. Industry is also looking very closely at criteria for citing new hazardous waste management facilities. CMA has developed the Hazardous Waste Response Center to track Superfund site clean ups, inventory and evaluate the technical remedies used at Superfund cleanups and advise member companies on a variety of Superfund related activities.

Lastly, several CMA member companies have funded a special study group to look critically at the issue of Underground Injection of wastes. The group has commissioned an in depth review of this controversial methodology. The results of that review will be published, along with certain layman level educational materials.

Ground water management concerns all of us. The purpose of this paper was to provide you with some background information, a perspective on the issue and the view of one segment of society that uses ground water.

18

GROUNDWATER CONTAMINATION

AN ENVIRONMENTALIST STRATEGY FOR GROUNDWATER PROTECTION

Brent Blackwelder

Dr. Brent Blackwelder is the director of the Water Resources Policy Project for the Environmental Policy Institute.

Points to Consider

1. What are some major faults with the EPA groundwater protection strategy?
2. Why does enforcement need drastic improvement?
3. What special strategies are recommended for the oil and chemical industries?
4. Why are budget requests by the EPA inadequate?

Excerpted from a groundwater policy statement by the Environmental Policy Institute, April 12, 1984.

Adequate water quality standards are essential because they provide the base limit on which to monitor and prosecute violators.

At the first national citizens conference on groundwater sponsored by EPI and 14 conservation, labor, and urban organizations, some 200 citizens from 40 states came to Washington to analyze and develop ideas for dealing with the growing problems of groundwater contamination. There was such a strong feeling about the inadequacy of EPA's groundwater strategy document that the Conference resolved to formulate our own far more comprehensive strategy. Here then is a summary of some components of a groundwater strategy as it is now taking shape.
. . .

1. **Improve Coordination of EPA Programs**—EPA has no coordination between setting standards under one program and correcting for those toxic requirements in other programs it regulates. Thus, the Toxic Substances Control Act (TSCA) enacted in 1976 regulates 53 substances which were tested and determined to be hazardous to human health. EPA under the Safe Drinking Water Act only regulates maximum contaminant levels of 21 substances. EPA has no apparent mechanism for forwarding the results of research from TSCA to incorporate in the Safe Drinking Water Act standards. This is just one example. There is little coordination between TSCA and the effluent guidelines established for industries under the Clean Water Act. The Resource Conservation and Recovery Act (RCRA) regulates 13 more substances than are controlled under the Clean Water Act. There is no coordination between substances controlled in the water and airborn emissions under the Clean Air Act, such as benzene. . . .

Under such obvious lack of rudimentary coordination, it is clear that groundwater protection cannot be achieved and public health cannot be protected.

2. **Enforcement Needs Drastic Improvement**—We recommend that the Congress and each state enact legislation to reward citizens who report pollution incidents leading to convictions with half of the fines levied by the court. Such legislation is now pending in Florida and is modeled after the state's Crimestoppers bill which provides rewards for citizens offering evidence to convict criminals. Such legislation if enacted would instantly provide a huge army of volunteers ready to enforce existing pollution laws.

The reason for suggesting this approach can be seen in the stunning lack of enforcement at EPA. Forty-five percent of the NPDES permits under the Clean Water Act have expired and there were less than four enforcement actions taken under the

121

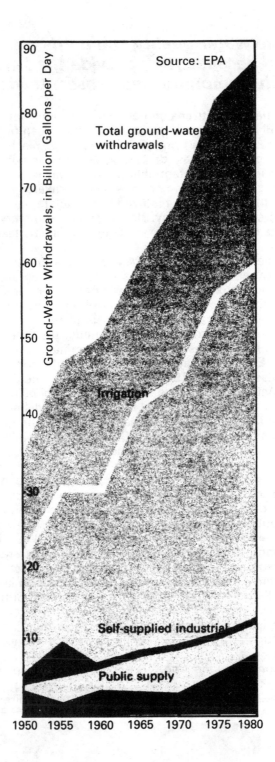

Source: EPA

Total ground-water withdrawals

Irrigation

Self-supplied industrial

Public supply

Act last year. RCRA regulates more than 30,000 sites that need permits, yet federal inspections of hazardous waste facilities decreased from 1825 to 1006 last year. It is estimated that 64% of these sites are now leaking but not one RCRA enforcement action has been undertaken.

It is essential for a radical improvement in inspections which serve as the basis for enforcement. A schedule must be established with adequate personnel so that all manufacturers and dischargers of pollutants can be inspected annually.

Adequate water quality standards are essential because they provide the base limit on which to monitor and prosecute violators. The Counsel of Europe and many individual states control more toxic limits in their water than EPA regulates nationally. Maximum contaminant levels for drinking water need to be greatly improved. The export and import of chemicals has increased by almost 25% since 1979. There are over 40,000 chemicals produced or in use in the United States; and 10–20 new chemicals enter into the market each week. However, TSCA has only added one new chemical, a standard for PCB, to the water quality standards.

3. **Mapping Requirements: a much more comprehensive and systematic program should be carried out by the U.S. Geological Survey**—The U.S. Geological Survey (USGS) was given a 5-year grant by Congress to study the interaction, degradation, and dispersion rates of a wide number of groundwater pollutants. This research is desirable and needs to be expanded, but going beyond this analysis it is essential that the United States develop better data on the boundaries of aquifers and recharge areas. Maps are the basis of many types of controls to safeguard against groundwater contamination and furthermore provide the background for monitoring and enforcing against violations. ...

4. **A comprehensive effort must be made to reduce the volume of waste generated and to recycle and reuse what is generated.**—There are numerous examples of industries looking more closely at a waste product and deciding that it can be turned into a profit through recycling or by selling it to someone else. Thirty-five non-profit waste exchanges have been organized across the country to identify and bring together industrial waste generators and potential users. These exchanges should be encouraged and perhaps a national 'commodities market' for wastes should be established. ...

5. **Special Strategies for the Oil and Chemical Industries**— The chemical and petroleum industries produce the preponderant share of toxic wastes which must be disposed of.

Efforts could be organized in major chemical producing states to analyze major polluting plants and promote source reduction. INFORM, a non-profit organization, has published a *Directory of 84 Organic Chemical Plants in California, a Directory of 95 Or-*

ganic *Chemical Plants in Ohio,* and a *Directory of Organic Chemical Plants in New Jersey.* These directories and a forthcoming report provide information on the size of the facilities, sources of their raw materials, types of products, amount of toxic wastes generated, and methods of reducing pollution at each plant.

The petroleum industry is not now regulated by EPA for underground storage tanks. A mandatory deadline for replacing old underground storage tanks must be passed. . . .

6. **Water Conservation Programs**—Comprehensive water conservation measures should be undertaken to increase the efficiency of water use in the municipal, agricultural and industrial sectors. Efficiency improvements will help minimize the amount of pollution potentially reaching groundwater.

7. **Federal Facilities should comply with groundwater protection laws**—Disclosures during the past year about the contamination of groundwater by federal facilities at Oak Ridge, Tennessee, and at the Savannah River Plant in South Carolina make it evident that the Federal Government can be and has been a major polluter of groundwater. There is a strong feeling that a repeat of previous experience with lawless federal agencies cannot be tolerated. The Department of Energy is perhaps the most serious problem.

8. **Increased Protection for Sole Source Aquifers**—EPA has a hostile attitude toward sole source aquifers, but citizens who have worked to obtain a sole source designation believe strongly that significant advantages have followed, including the passage of state legislation helping to protect the designated aquifer.

There should be increased designation of sole source aquifers, and the protections afforded these aquifers should be strengthened.

9. **Adequate Budget**—Although an Office of Groundwater has

been created at EPA, there is no line item budget request for the Office for FY 1984 or FY 1985. It is essential that this office be properly funded and we recommend $1.5 million for a supplemental 1984 budget and $3 million for fiscal 1985. In addition, more money is needed for adequate inspection and enforcement teams, and a slot for water conservation should be properly funded. . . .

10. **Underground Injection Control (UIC)**—Over 221,000 wells in the U.S. inject liquid wastes into the earth. As environmental regulations in the 1970's began to limit disposal of wastes into the air, water and landfills, underground injection became more widespread. Now EPA studies suggest that underground injection has overtaken land disposal by a ratio of 2 to 1. . . .

States need to rapidly improve the data they have on the number of wells now in existence; and to require accurate inventory from each injection well on the type and amount of substances injected. Two states, Florida and Alabama, have banned all UIC activities because of the geological formation in their area.

The use of incorrect drilling procedures and improper rates of injection have caused groundwater problems. States should establish strict controls over the drilling technology used, monitor the wells during the drilling and operational phase, and limit the types of substances which can be injected.

11. **Pesticide Contamination of Groundwater**—As the problem of pesticide contamination of groundwater is becoming publicized, some action is being taken. The State of Wisconsin is acting to attempt to zone to restrict the use of a pesticide in certain areas where groundwater monitoring can identify that the soil can not trap and degrade the toxic chemicals but rather would enter the groundwater.

A short-range strategy to protect groundwater from pesticide contamination would involve the enactment of amendments to the Federal Insecticide, Fungicide and Rodenticide Act to restrict the use of water-soluble agricultural chemicals in aquifer-recharge areas. . . .

12. **Monitoring**—While EPA asserts that a national ambient monitoring program would not be feasible much can be done to improve coordination between existing regulatory programs. Groundwater monitoring collected for Superfund sites, RCRA waste disposal facilities, special studies conducted under the Clean Water Act, and proposed monitoring of underground storage tanks under TSCA needs to be brought together in one central data bank. . . .

Requiring public water suppliers to test their drinking water for the most common industrial chemicals and wastes and pesticides used within the watershed would provide an important safeguard for consumers.

THE CLEAN SITES
ALTERNATIVE

Charles W. Powers and the Conservation Foundation

Charles W. Powers is the President of Clean Sites, a non-profit corporation organized to hasten the cleanup of hazardous waste sites that threaten groundwater supplies and pose other serious problems. It attempts to coordinate the resources of private companies, government agencies, and environmental groups. It encourages major efforts and leadership in the private sector for hazardous waste cleanup. Clean sites was organized by the Conservation Foundation in Washington, D.C.

Points to Consider

1. How is Superfund defined and what has it accomplished?
2. What are the major goals of Clean Sites?
3. Why is the private sector an important element in hazardous waste cleanup?
4. What specific steps does Clean Sites take to help clean up hazardous waste dumps?

Excerpted from a 1984 pamphlet on hazardous waste by the Conservation Foundation, and Charles W. Powers, "The Idea Behind Clean Sites," *EPA Journal*, October, 1984.

Clean Sites believes that in very many cases the best resource is the private sector.

Concern about groundwater pollution has grown substantially in recent years as we have learned more about the extent and potential seriousness of the problem. The sources of contamination are diverse: they include hazardous waste dumps, surface impoundments for liquid wastes, septic tanks, underground storage tanks, salt water intrusion, deep well injection, mine tailings, municipal waste systems, and pesticides and fertilizers applied to cropland.

There is very little monitoring of groundwater quality, and no federal law is primarily concerned with preventing groundwater pollution (several EPA laws do have provisions which directly or indirectly relate to groundwater). Although some states have enacted groundwater protection statutes, there has traditionally been much less attention paid to groundwater management (both quantity and quality) than to surface water management.

RCRA: The First Response

As stories of "midnight dumpers" multiplied, Congress took a first major step to address the problem of waste generation and disposal. In 1976, it passed the Resource Conservation and Recovery Act (RCRA). Open dumping was largely stopped; proper solid waste management and resource recovery practices began to develop; EPA began to provide technical and financial assistance to state and local governments that wanted to grapple with the problem. Finally, EPA began to regulate the treatment, storage, transportation, and disposal of hazardous and non-hazardous wastes that seemed to have the potential to pose adverse effects on health and the environment.

The Problem Festers: "Old Sites" and Groundwater Contamination

If society believed that it had adequately dealt with the problem of waste disposal through RCRA, Love Canal taught a different lesson. EPA tried with limited success to use its existing legal authority to respond to the growing knowledge that there might be a significant number of these sites. Where the government could find parties responsible for creating the conditions needing cleanup, it had no clear legal authority to establish liabilities for these past practices. Moreover, many of the parties responsible for the dumps could not even be found and those who could be found frequently did not have the money to pay for cleanups.

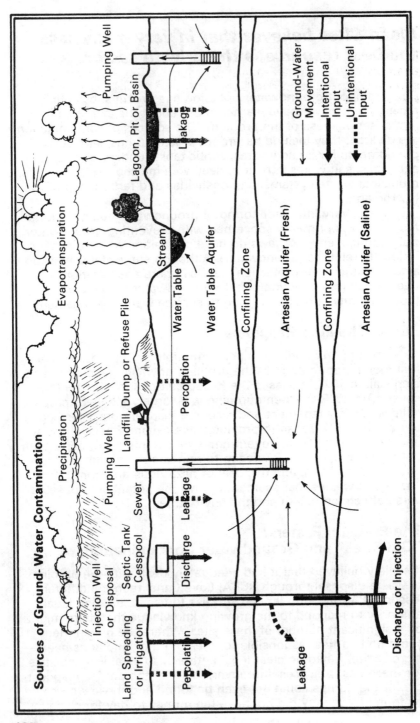

Sources of Ground-Water Contamination

Source: EPA

128

The Enactment of Superfund

In response, Congress tackled the hazardous waste problem again in 1980. It passed the Comprehensive Environmental Response Compensation and Liability Act (CERCLA), commonly known as "Superfund." The law assists EPA in two major ways: (1) it provides *funding* for the government to clean up inactive waste sites where responsible parties cannot be found or are unable or unwilling to perform the cleanup; (2) it creates *liabilities* for parties who were associated with waste sites, either to perform the cleanup work or to reimburse EPA for its costs.

Funding. Seven-eighths of the Superfund dollars come from a special tax on the sale of petrochemical feedstocks, crude oil, and other chemicals. The rest comes from general federal revenues. The fund was projected to reach $1.6 billion in a five-year period ending in September 1985.

Liabilities. Companies or individuals with certain relationships to a problem site may be liable for the costs of cleaning up the site. EPA also has the authority under some circumstances to compel parties to perform the cleanup work. Potentially liable parties include:

- the site owners or operators,
- transporters of waste to the site, and
- generators who arranged for their waste to be sent to the site.

The structure of Superfund provides EPA with two basic options in any site cleanup:

- It encourages EPA to seek private action through the use of voluntary agreements and administrative or judicial orders.
- It authorizes EPA and the U.S. Army Corps of Engineers to perform work on its own and seek reimbursement from responsible parties.

In any place where the government performs cleanup on its own with Superfund money, the state in which the site is located must contribute 10 percent of the capital costs of cleanup and must pay for all operation and maintenance expenses. If the site is owned by the state or by a political subdivision of the state, the state must contribute at least 50 percent of the capital costs of cleanup. . . .

The Problem Remains

Three years after the passage of Superfund, only six sites have been cleaned up through the public funds process. This program is growing, but the pace must be accelerated, particularly because the potential consequences of delay now seem great. Point source leakage at some sites is now a reality. While legitimate disagreements occur concerning the migration of chemicals into the groundwater at some sites, the fact is that inaction will only exacerbate problems.

Even if the government can undertake all the actions it envisions by the end of fiscal year 1985, it will deal with only a fraction of the sites on the National Priority List. Even with a reauthorization of Superfund, a sizable component of the priority sites will probably not be reached for a very long time.

The problem is *not* inherently unmanageable. Instead of operating from the current list of 16,000 sites, a major effort on 1,000 to 2,200 sites could substantially solve the problem. How to accomplish that by increased private sector action is the problem with which we have wrestled for the past several months. . . .

Harnessing the Private Sector

Clean Sites, Inc., is a new nonprofit corporation formed on May 31, 1984, "specifically to encourage, contribute to, and bring about the cleanup of hazardous waste sites in the United States." Its Board of Directors includes two former EPA Administrators (one of whom now heads a major environmental organization), two presidents of major universities, three chief executive officers of major U.S. firms, a leading state official, the head of a major conservation research foundation, and the head of an environmental health research organization. . . .

Clean Sites is not an alternative to a strong EPA program, but an additional resource. Its purpose is to help speed cleanup activity by providing resources and skills at the points where existing institutions are limited in what they can do and where complex agreements often get stuck. . . .

Clean Sites believes that in very many cases the best resource is the private sector. When it comes to developing solutions to

130

the problems at sites—even multi-party sites—the generators probably know most about the substances and how to treat them without increased risk to cleanup personnel and local citizens. Generators, along with transporters and owners, probably know most about what is at the site and how it got there. The researchers and engineers who work for the generators probably have the most technical expertise. And finally, the responsible parties provide an important set of resources, particularly management resources, to augment those available in EPA and other governmental organizations. In the battle against hazardous sites, these capabilities are crucial.

Private sector cleanup, however, poses tough issues about how to coordinate with government oversight and how to determine financial responsibility. Clean Sites was created for the sole purpose of addressing these issues and fostering the process of agreement at those sites where effective and rapid cleanup can best be achieved by private parties.

Clean Sites hopes to do this by providing resources at a number of steps in the process, particularly at the transition points where:

- the effort to reach agreements between responsible parties themselves and with the government and other publics traditionally breaks down;
- there is a need for third-party, independent help in apportioning responsibility in difficult cases;
- government needs to be assured that proposed settlements are worthy of commitment of its scarce human resources;
- a mechanism is needed to allow public and private funds to be mixed to secure a better and more rapid cleanup.

A Third Party

The ten-month Conservation Foundation effort which led to Clean Sites' creation pinpointed the need for a third-party institution. The Clean Sites Board of Directors has been painstakingly chosen to engage a group of Americans who can ensure commitment to environmental protection; foster innovative ways to keep affected publics informed and to seek their views and their acceptance; provide resources for technical competence; and spur creativity and effectiveness in the development and acceptance of site remedies.

RECOGNIZING AUTHOR'S POINT OF VIEW

This activity may be used as an individualized study guide for students in libraries and resource centers or as a discussion catalyst in small group and classroom discussions.

Many readers are unaware that written material usually expresses an opinion or bias. The capacity to recognize an author's point of view is an essential reading skill. The skill to read with insight and understanding involves the ability to detect different kinds of opinions or bias. Sex bias, race bias, ethnocentric bias, political bias and religious bias are five basic kinds of opinions expressed in editorials and all literature that attempts to persuade. They are briefly defined in the glossary below.

FIVE KINDS OF EDITORIAL OPINION OR BIAS

sex bias—the expression of dislike for and/or feeling of superiority over the opposite sex or a particular sexual minority

race bias—the expression of dislike for and/or feeling of superiority over a racial group

ethnocentric bias—the expression of a belief that one's own group, race, religion, culture or nation is superior. Ethnocentric persons judge others by their own standards and values

political bias—the expression of political opinions and attitudes about domestic or foreign affairs

religious bias—the expression of a religious belief or attitude

Guidelines

1. Locate three examples of political opinion or bias in the readings from chapter three.

2. Locate five sentences that provide examples of any kind of editorial opinion or bias from the readings in chapter three.

132

3. Write down each of the sentences referred to in guideline two and determine what kind of bias each sentence represents. Is it **sex bias, race bias, ethnocentric bias, political bias,** or **religious bias?**

4. Make up one sentence statements that would be an example of each of the following: **sex bias, race bias, ethnocentric bias, political bias** and **religious bias.**

5. See if you can locate five sentences that are factual statements from the readings in chapter three.

6. Summarize the author's point of view in one sentence for each of the readings in chapter three of this book:

Reading 14_____

_____

Reading 15_____

_____

Reading 16_____

_____

Reading 17_____

_____

Reading 18_____

_____

Reading 19_____

_____

CHAPTER 4

PROTECTING SURFACE WATER

READINGS

20

PROTECTING SURFACE WATER

NONPOINT POLLUTION SOURCES: AN OVERVIEW

Edwin H. Clark II

Edwin H. Clark is a senior research associate at the Conservation Foundation, a not for profit environmental policy research organization. He is a specialist in soil erosion.

Points to Consider

1. How extensive is agricultural erosion?
2. What kind of pollution problems does it pose?
3. What are the economic costs of nonpoint source pollution?
4. How much pollution in our waterways is nonpoint pollution responsible for?

Reprinted from a public statement by the Conservation Foundation, July, 1984.

135

The total economic costs caused by sediment and pollutants associated with erosion amount to $2.5 to $3.0 billion dollars a year.

Our failure to control nonpoint source pollution is imposing substantial economic costs on the country, perhaps measured in the billions of dollars. In order to provide a better feeling for the potential significance of this source of water pollution, it may be useful first to look at some of the basic data on soil erosion in the United States.

In 1977, 6.4 billion tons of soil eroded from non-federal lands, above and beyond normal geological erosion. This is equivalent to about 30 tons of soil for each person in the United States, or to an average of more than 220 tons of erosion every second of the year. About 1.5 billion tons of this was wind erosion from cropland and rangeland in the Great Plains states. The rest was water erosion of cropland (39%), pastureland and rangeland (30%), and a variety of other sources including forestland, streambanks, gulleys, roads, and construction sites.

Some simple comparisons on agricultural erosion are instructive. For instance, the State of Iowa is famous for its corn production. But it also produces a lot of erosion. In fact, for every one ton of corn raised by an Iowa farmer in 1977, five tons of soil were eroded by water from that state's cropland. Nationally, for each pound of food consumed in the United States, over 22 pounds of soil were eroded from all agricultural lands by water. The average rate of water erosion of cropland alone was more than 89 tons per second.

Much of this erosion comes from a relatively small portion of the land. This is an important fact to remember when considering the need for erosion control measures. For instance, less than 3 percent of the land, eroding at a rate of more than 25 tons per acre per year, is producing almost a third of the total erosion (and even a higher proportion, almost 60 percent, of the "excess" erosion—this is the erosion in excess of 5 tons per acre which is [somewhat arbitrarily] considered to be safe).

The majority of this eroded soil eventually reaches the nation's waterways, and once there can cause serious problems. It can destroy valuable fish populations in streams and bays, increase the cost of water treatment for downstream users, make valuable recreational facilities essentially useless, increase significantly the flood damages along rivers, interfere with navigation, and silt up valuable storage space in lakes and reservoirs.

Other, perhaps more serious problems arise because erosion from cropland not only includes natural soil, but other contaminants such as fertilizers, pesticides and oxygen consuming organic materials. Pesticides can be toxic to fish, and potentially to humans. Fertilizers can be a source of potentially toxic nitrites in drinking water supplies, and stimulate algae growth, accelerating the eutrophication of lakes and reservoirs. As algae die and decay, they, along with the plant residues and other organic materials carried into the stream, use up oxygen, often reducing the oxygen level so much that some fish species can no longer survive. These off-site pollution problems are worsened when farmers try to compensate for lost productivity by using more fertilizers, pesticides, and other inputs, without reducing the amount of erosion.

The Economic Costs

The total economic costs caused by sediment and pollutants associated with erosion amount to $2.5 to $3.0 billion dollars a year. Some of these costs—15 to 20 percent—result from impacts commonly associated with water pollution; for instance, $40 million for increased water treatment costs, another $40 million for damages to inland commercial fisheries, and $100 to $500 million in damages to recreational opportunities.

But many of the costs are of a different type, and may not normally be linked to water pollution problems. For instance, about 25 percent are related to dredging. The Corps of Engineers spent $362 million in 1980 dredging harbors and navigation channels. The purpose of the dredging is to remove sediment deposits, and sediment comes from soil erosion. These federal expenditures are thought to be matched by like amounts spent for dredging by state and local governments and private firms.

The sediment is also filling up our water supply and flood control reservoirs, and this accounts for about 30 percent of the costs. Twenty percent of the capacity of new reservoirs—costing about $300 million a year—is built for the sole purpose of storing sediment. An estimated one million acre feet of existing reservoir capacity is being filled each year by sediment. That capacity has to be replaced, or someone will be short of water. Even if we could replace it at 1980 costs, the bill would come to some $300 million. In flood control projects and other small reservoirs, the project managers are able to remove the sediment—and probably pay $200 million a year doing so. And finally, reservoir and lake owners are spending significant amounts—an estimated $100 million a year—to attempt to control algae, seaweed and other problems caused by nutrients which are associated with soil erosion.

The other major cost area is flood damages, accounting for 20

Source: EPA

percent of the costs. Sediment-clogged channels flood more
often and the floods are bigger than they would be if there were
no erosion. Even more important is the damage caused by sediment—a major portion of urban flood damages are not caused
by the water, but rather the sediment and other pollutants it
leaves behind. And sediment deposits also destroy millions of
dollars worth of agricultural products every year and reduce the
long-term productivity of the flooded land. The total costs of

these sediment-related damages are probably about $600 million annually.

The final type of cost is the cost of removing sediment from drainage ditches and irrigation canals. These channel maintenance costs amount to an estimated $200 million per year.

Again, the sum of all these preliminary estimates is $2.5 to $3.0 billion a year. And this does not include many impacts for which there is not enough quantitative information to make even crude cost estimates.

In addition to emphasizing that these estimates are still preliminary, I should also point out that even the most complete soil conservation or nonpoint source pollution control program would not result in these costs being entirely eliminated. Many of them are caused, at least in part, by "natural" erosion processes that can never be completely controlled.

Even so, I believe studies demonstrate the very high costs that the nation is experiencing because it has not yet attempted to control nonpoint sources of pollution. If we are serious about attaining our goals of fishable, swimmable waters, we are clearly going to have to do something about these sources.

Conclusion

According to one recent estimate, nonpoint sources are currently responsible for about three-fourths of the sediment in our rivers, over 90 percent of fecal and other coliform bacteria, 80 percent of the total nitrogen and 50 percent of the phosphorus that reach our waterways. They are the major source of pesticide pollution, and an important source of other toxic pollutants. For instance, over 90 percent of the lead reaching most of the Great Lakes comes from nonpoint sources.

In some cases the impact of these sources is more serious than even the total amount of pollution would indicate because their discharge is concentrated during storm periods. Although the increased water in the rivers associated with a storm provides a dilution effect, the concentration of pollutants, for instance of sediment, will be much higher following a storm than during normal times. And the run off from a city's streets with a summer storm may cause a much more serious impact on oxygen levels in the receiving stream than the steady contribution of the municipal sewage plant.

A NATIONAL NONPOINT POLLUTION STRATEGY

Robert J. Johnson

Robert J. Johnson is president of the North American Lake Management Society (NALMS), a national organization made up of citizens, scientists, lake associations and water pollution control professionals who are concerned about poor water quality in our nation's lakes.

Points to Consider

1. What is the national scope of lake pollution?
2. How are the goals of the Clean Water Act defined?
3. What are some Best Management Practices (BMP's) that have been effective?
4. Why is a national nonpoint source program needed?

Reprinted from a public statement by the North American Lake Management Society, July 19, 1983.

It is said that the Missouri basin, with its high density of livestock operations, has an organic pollution wasteload equivalent to the population of China.

The lakes and impoundments of this Nation are vital not only for their ecological functions in supporting fish and wildlife but also for the economic basis—measured in the billions of dollars—to many local, state, and national businesses. Unfortunately, lakes are very susceptible to pollution because of their ability to capture pollutants and accumulate them within their water, bottom sediments, and aquatic life. . . .

It is less cost effective to clean up lakes and rivers after they get polluted than to prevent them from being degraded. The longer Congress delays in addressing priority nonpoint source problems, the more expensive it will be for our children and grandchildren to restore suitable quality to our Nation's waters. And the more difficult that task of restoration becomes. . . .

Flood damage and dredging costs are extremely high. Certainly several hundreds of millions of dollars have been expended to keep waters open for navigation over the last five years. Federal water projects such as Fishtrap and Dewey Reservoirs in Kentucky are filling with sediment. In the West, sediment is deposited in Federal water projects and displaces water that could be stored for irrigation. The state of Wyoming has identified Boysen Lake as such an example suffering from accelerated sedimentation. . . .

The public is very concerned about contamination of its recreational areas, its food resources, and its drinking water. Urban runoff and sewer leakages have caused public health concerns about bacterial contamination across the country.

Fish are contaminated in urban areas as well as rural. . . .

Meeting Clean Water Act Goals

The interim goal of the Clean Water Act involves ensuring a water quality for the protection and propagation of aquatic life and recreation in and on the water. Unfortunately, this goal is not being met across the country in rivers, streams, lakes, and estuaries and not only imposes social costs on the public, but it also causes actual economic damage. . . .

Nonpoint source pollutants from agricultural activities and urban areas cause the majority of nonpoint problems in our Nation. It is said that the Missouri basin, with its high density of livestock operations, has an organic pollution wasteload equivalent to the population of China. Some of these operations are classified as point sources, many are classified as nonpoint sources. Excessive loadings of nutrients from animal waste and cropland have caused serious eutrophication previously mentioned in lakes (algal blooms, weeds, fishkills) that concern the public. . . .

Do Best Management Practices Work?

It is clear that economic damage of national importance is being caused by nonpoint sources. The public is concerned about health implications of water quality problems caused by these diffuse sources, and goals of the Clean Water Act are not being attained across the Nation because of this pollution. Similar to the debate over acid precipitation, opponents of BMP (best management practices) implementation say that not enough information is known about cause and effect, cost effectiveness, or social acceptability of BMPs. I would answer these concerns by saying that the nonpoint problems are known and are causing unacceptable economic damage, scientists have shown BMPs to be effective in reducing nonpoint source loads, and it is time to simply implement what we know before our valuable lakes and estuaries are irretrievably lost.

The EPA Clean Lakes Program was a leader in years past in funding implementation of urban and agricultural best management practices to protect lake quality. Enough experience has been gained from that program to know that BMPs are effective in protecting water quality. Diversion of the first flush of urban stormwater to recharge basins was effective at Lake Eola, Florida. Wetlands can even be created to filter urban runoff in certain areas. This was done at Lake Jackson, Florida. Retention basins and wetlands were utilized effectively to remove urban stormwater pollutants at Clear Lake and Phalen Lake in Minnesota. Diversion of stormwater at Mirror Lake in Wisconsin reduced phosphorus concentrations 65 percent and algal blooms did not recur. Settling ponds were effective at Hylan Lake, Minnesota, and Temescal Lake, California. The EPA Nationwide Urban Runoff Program is finding detention basins to be quite effective in improving stormwater quality. Suspended solids reductions commonly exceed 70 percent, nutrient reductions exceed 50 percent, and heavy metals reductions exceed 70 percent. BMPs have also been developed to abate combined sewer overflow problems. The BMP program for Syracuse, New York to protect Onandaga Lake is a good example of cost-effective sewer overflow abatement to protect lake quality.

Agricultural BMPs have also been implemented under EPA's Clean Lakes Program. Agricultural BMP implementation has

142

The Major Polluter

Non-point source pollution provides approximately one-half of all the conventional water pollution problems in America. In many cases, non-point source contributions from agriculture, urban areas, forestry, and construction sites overwhelm the efforts made to date to address the point source contributors.

In other words, the millions of dollars spent to clean up point sources are being overwhelmed by non-point source pollution. Here are some statistics: 75% of all of our sediment, 90% of fecal and other coliforms, 80% of nitrogen, 50% of phosphorus, a major cause of pesticide pollution. In other words, the pesticides would stay where they are and largely break down in the soils were it not for loss of soils to waterways and then converting into serious pesticide pollution problems. Also, our non-point source problem is a major cause or source of biological oxygen demand.

Izaak Walton League of America, 1983

been funded for agricultural pollution control in the watersheds of Spiritwood Lake (North Dakota), Cochrane Lake and Herman Lake (South Dakota), Cobbossee Lake (Maine), Skinner Lake (Wisconsin), Broadway Lake (South Carolina), and Rivanna Lake (Virginia). At White Clay Lake, BMPs resulted in about 60 percent reductions in annual phosphorus loading. Results are also available from USDA's Model Implementation Program (MIP). BMPs helped to reduce nutrient and sediment loads at Cannonsville Reservoir (New York), Lake Herman (South Dakota), and Pittsfield Lake (Illinois). . . .

A National Nonpoint Source Program

Many members of the Society deal daily with lake management problems. They have found that BMPs were not extensively implemented as part of the voluntary plans prepared as part of the 208 program. The 208 plans are insufficient, and in some cases, states did not declare BMPs, lest BMP implementation be required. It is the Society's observation that BMPs can be implemented, for example as part of EPA's Clean Lakes Program, when the government pays for part of them (cost sharing or other public subsidies). There are still many land users in these projects that do not wish to comply with BMP implementation and there are questions as to whether sufficient water quality improvement will take place if these key land users do not participate.

A national program to control nonpoint sources is desperately needed to mitigate economic damage caused by the pollution, to restore designated uses to waters, to protect water quality for future generations, and to preserve valuable land and water resources. In general, the Society supports the national nonpoint source program. We especially support the need to base identification of waters adversely affected by nonpoint sources not only on violations of applicable standards but also on designated uses of the waters that are not attained because of nonpoint pollution. This is an important requirement because many problems that degrade lake and estuarine water quality—such as sedimentation and symptoms of eutrophication—do not have applicable standards established by the states. We also support the provision for a commitment in identifying strategies of how management practices and control measures will be implemented to restore and maintain water quality suitable to attain the goals of the Act within a specific timeframe. The specific timeframe and a commitment on strategy is necessary to avoid repeating earlier mistakes. The 50 percent matching requirement for Federal fund eligibility for implementation is a valuable element of the program. Federal, State, and local cooperation is essential to restore waters degraded by nonpoint sources and this matching requirement will assure—as it does in the EPA Clean Lakes Program—that state and local commitments to implementation will be made since they will devote their funding to the program.

Interstate Pollution

NALMS is also concerned that the proposed provision regarding interstate nonpoint source pollution may be insufficient to achieve attainment and maintenance of applicable standards, designated uses, or goals of the Act. This interstate pollution causes priority problems across the country, from the Snake and Colorado Rivers to Palisades Reservoir (Idaho and Wyoming), Big Stone Lake (South Dakota and Minnesota), the Great Lakes, and the Chowan River (Virginia and North Carolina). The Federal government should take the lead in resolving problems of an interstate nature if that water does not meet the goal of the Clean Water Act. A more aggressive Federal role than the one proposed in proposed section 11 will be required to meet citizens' concerns about waters not attaining the goals of the Act.

NALMS supports the basic thrust of a National Program. . . . It will encourage cooperative, interagency approaches to correcting existing problem areas to protect water quality, and to preserve valuable land resources for sustained use. The Society wishes to reiterate that without strong Federal leadership, specific milestone dates to measure progress, and public participation, nonpoint sources will continue to cause economic damage across the country and lakes as well as estuarine resources may be irretrievably lost.

A STATE CENTERED NONPOINT PROGRAM

American Paper Institute (API) and The National Forest Products Association (NFPA)

API/NFPA are both national trade organizations representing over 2,000 member companies who own and manage forest lands, manufacture pulp, paper and solid wood products.

Points to Consider

1. Why are state and local government agencies better able to implement nonpoint source pollution controls?
2. What kind of federal financial support is needed?
3. What is the appropriate federal role?
4. When should financial sanctions and penalties be used?

Excerpted from a position paper by the American Paper Institute and the National Forest Products Association, 1983.

First, the threat of direct federal intervention to implement nonpoint source controls rings hollow in light of the lack of federal capability to perform this function.

The nature and variability of nonpoint source problems and needed control measures dictate that control programs must be developed and implemented at the state and local level. In our view, there are very real institutional and resource limitations that would be faced by EPA or any other federal agency attempting to implement, monitor, and enforce nonpoint source controls at the state level. These limitations make direct federal implementation of nonpoint source control programs infeasible, even in cases where states are, for some reason, unwilling or unable to act.

Quite apart from the issue of the federal government's questionable capability to move to implement a program in place of a state, we wonder whether speculation over the possible inaction of a state or states is a real concern. The survey by the National Council of the Paper Industry for Air and Stream Improvement (NCASI) of state silviculture nonpoint source pollution control programs indicates that forty states are moving forward in the absence of any federal assistance whatsoever. The recent survey conducted by the National Association of Conservation Districts demonstrates a similar degree of state and local commitment in the agricultural area.

The principal impediment to more aggressive efforts to control nonpoint sources over the past three years has not been a lack of state or local interest. Rather, efforts have slowed to some extent because of a perceptible lack of federal support, encouragement, and/or financial assistance to augment state and local efforts.

In this light, we believe it is, at best, premature to threaten heavy-handed federal intervention or sanctions in states which for some reason fail to move forward as fast as many would like. First, the threat of direct federal intervention to implement nonpoint source controls rings hollow in light of the lack of federal capability to perform this function. Second, it is not at all clear that this is the only, or even the best, means of inducing reluctant states—if there are any—to move forward in an expeditious fashion.

API/NFPA suggest an alternative role for EPA in this process. We believe that the Administrator should review the status of nonpoint source problems in those states which fail to submit

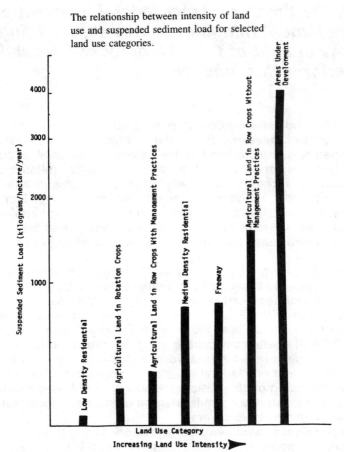

The relationship between intensity of land use and suspended sediment load for selected land use categories.

Source: Wisconsin Department of Natural Resources

an implementation plan under whatever proposal is finally adopted. The Administrator should report to Congress within eighteen months assessing: (1) the need for additional nonpoint source controls in these states; and (2) appropriate federal sanctions and remedies as a result of the states' failure to act.

Is Federal Financial Support Necessary to Foster the Implementation of Nonpoint Source Controls?

Our industry survey of the current status of state silviculture nonpoint source control programs indicated that a major impediment to more rapid implementation of nonpoint source controls is a lack of available financial and staff resources at the state

148

level. Indeed, ten of the forty states which are actively implementing a silviculture nonpoint source control program responded that lack of federal financial support, as Section 208
grants are phased-out, will slow down their efforts.

Major federal financial assistance to the states for dealing
with nonpoint source problems ended in 1980 with the completion of the Section 208 planning process. Since that time, there
have been limited federal financial measures provided to the
states to help implement specific Section 208 plan elements. In
our view, given the current limitations on resources at the state
and local level, federal financial assistance will probably be necessary. . . .

We appreciate the current lack of hard information concerning
the total financial resource commitment necessary to effectively
address nonpoint source controls nationwide in order to achieve
the goals of the Act. At the same time, we believe that a larger
federal commitment, at the present time, would be helpful to begin to move toward this end. . . .

Beyond Financial Support, What is the Appropriate Federal Role in Addressing Nonpoint Source Pollution?

API/NFPA believe that the implementation of nonpoint source
control programs should remain a state and local responsibility.
However, there is an appropriate role (in addition to providing
financial assistance) for the federal government in this process.

The Environmental Protection Agency should continue to review and approve state nonpoint source control programs in order to: (1) assure reasonable state progress toward meeting
water quality goals; (2) guard against unwise expenditure of federal financial resources in supporting state programs that are
poorly conceived and/or operated; and (3) resolve any interstate
disputes concerning nonpoint source problems. In our view, EPA
should also serve as a clearinghouse for the transfer of information on successful implementation programs at the state level.
Finally, EPA should assess the progress being made, and keep
Congress abreast of needed changes to the federal strategy for
dealing with nonpoint sources. This responsibility should include, as suggested above, a requirement for EPA to report to
Congress where states do not act to address nonpoint source
problems.

In addition to EPA, we believe that the U.S. Department of Agriculture has an important role to play in addressing nonpoint
source problems. USDA should provide technical assistance,
where appropriate, to assist state implementation of nonpoint
source control programs. USDA's expert agencies and delivery

systems for dealing with the agricultural community will prove essential in assisting the states to move forward in addressing water quality problems caused by agricultural activities.

Finally, we believe that all federal agencies have a responsibility to work closely with state nonpoint source control implementation agencies to make sure that federal programs are consistent with state efforts in this area. The federal government should not unwittingly cancel out the effects of state nonpoint source control efforts as the result of poorly coordinated federal projects or programs.

Should States Be Bound to a Uniform, Nationwide Date by Which Nonpoint Source Controls Must Be in Place?

Another important finding is that the variability of nonpoint source problems from state to state—and even within states in certain areas—requires a policy that provides the states with considerable flexibility in controlling nonpoint source pollution. In our view, the most important components of this policy include the flexibility to develop: (1) different types of control programs, both regulatory and non-regulatory; (2) schedules for achieving implementation milestones that reflect the necessary differences in these control programs; and (3) economically and technically feasible best management practices that will be applied in a site-specific fashion. . . .

The local variability of nonpoint source problems makes it imperative that states be afforded varying time schedules to implement necessary controls.

We believe that a uniform deadline for compliance could well have a chilling effect on the states' willingness and ability to consider a broad range of alternative implementation programs. Such a deadline runs a very real risk that the states will identify only those problems for which they are certain they can achieve compliance within the forty-eight month deadline.

150

API/NFPA prefer a less extreme option than attempting to impose nationwide nonpoint source controls within forty-eight months, an approach which appears to us to represent point source type thinking. We suggest that each state program submitted to EPA for approval include a schedule of accomplishments—agreed upon by both the state and the Administrator—for the implementation of the nonpoint source controls specified in the state plan. By monitoring the state's progress in achieving these milestones, we believe that EPA will have adequate opportunity to assure the state is moving forward on a timely basis, in a fashion that is reflective of each state's particular nonpoint source problems. This approach will result in a more honest appraisal of the nonpoint source problems in each state, as well as a more sincere effort to deal with these difficulties.

Can Financial Sanctions Be Used Effectively to Compel Compliance with Nonpoint Source Pollution Controls?

One suggestion introduced involves cutting off federal financial assistance for any activity not conducted in compliance with required nonpoint source pollution control practices. On the surface, this proposal appears appealing because it does not involve any additional federal financial exposure. Indeed, such an approach could work to reduce the amount of federal financial assistance provided through various federal support programs.

In our view, however, the appeal of such an approach is largely superficial. The element of compulsion injected by such proposals for financial sanctions conflicts with current EPA policy to foster cooperative programs to implement nonpoint source controls. Experience suggests that an approach which emphasizes education, training, and, where necessary, incentives, offers the best chance of success, especially in dealing with rural nonpoint source problems. A controversial sanctions program may make states reluctant to honestly identify problems and develop control programs.

As an alternative to this controversial approach, API/NFPA support efforts by federal agencies to work with states to assure that the federal programs are consistent with state nonpoint source control efforts. More generally, we believe that positive initiatives rather than financial sanctions against nonpoint sources will have the greatest impact on improving water quality.

ACID RAIN AND SURFACE WATERS: OVERVIEW

Akwesasne Notes & the GAO

The following comments were excerpted from Akwesasne Notes, *the official publication of the Mohawk Nation at Akwesasne and a Government Accounting Office report on acid rain.*

Points to Consider

1. How does acid rain effect lake chemistry?
2. Because of acid rain, how many lakes in the U.S., Canada and Sweden will no longer support life?
3. How many lakes in the U.S. are endangered by acid rain?
4. What control techniques can be used to reduce acid rain?

"Acid Rain: First Rumors of Ecocide," *Akwesasne Notes,* Winter, 1985, and *Analysis of Issues Concerning Acid Rain,* GAO, December 11, 1984.

Svante Oden's rain maps clearly showed the growing acidity spreading out from pollution sources in Europe's industrial center.

AKWESASNE NOTES

In the mid 1960's a Swedish scientist, Svante Oden, discovered something about the composition of Europe's rainfall which was to become the first portent of the price we shall eventually pay for widespread air pollution. When Svante Oden started mapping the chemistry of Europe's rainfall, an expanding circumference of increasing environmental acidification began to emerge. While America was enjoying unprecedented economic prosperity—the so-called American Golden Age of the 1950's, acidification of the environment was increasing dramatically. Although there were no visible symptoms of lake or vegetation sickness at that time, Svante Oden's rain maps clearly showed the growing acidity spreading out from pollution sources in Europe's industrial center.

Then, the Swedish scientist made an important connection between acidified atmospheric conditions to the fish kill which was occurring in western Sweden. At first ridiculed, the notion that rainfall could become acidified by industrial pollutants has been confirmed by many atmospheric scientists so that there no longer exists any doubt that Acid Rain is in fact a by-product of industrial activities and our hydrocarbon lifestyle. The fish kill in western Sweden was only a prelude to the subsequent death and damage to thousands of acidified lakes in the higher elevation areas in Europe and the Northeastern United States.

Lake Chemistry

To biologists who are monitoring the environmental effects of acid precipitation, the specter of lake decline has become a familiar pattern. When a lake becomes acidified, the fish and plant life die species by species until, with the exception of some algae and moss, no life can exist in the lake as it becomes severely acidified.

Lake chemistry, which is always closely linked to the land, is changed by the action of acid rain. The Eastern portion of the United States receives rain which is 10 times more acidic than what is considered normal. This Acid Rain dissolves aluminum from the soils surrounding a lake and enters the lake water. Fish are especially vulnerable to this combination of aluminum and acid, their gills erode, become clogged and the fish die from suf-

153

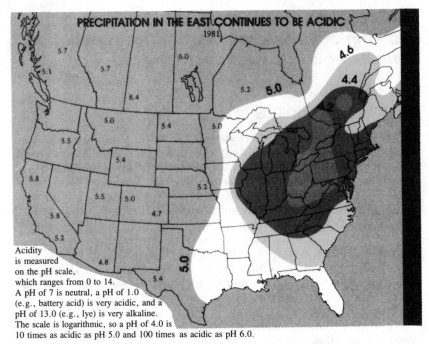

PRECIPITATION IN THE EAST CONTINUES TO BE ACIDIC
1981

5.7
6.0
4.6
5.1
5.7
4.4
6.4
5.2 5.0
5.0
5.4 5.0
5.5
5.4
5.8
5.2
5.5 5.0
5.8 4.7
5.2
Acidity
is measured 4.8
on the pH scale, 5.0
which ranges from 0 to 14. 5.4
A pH of 7 is neutral, a pH of 1.0
(e.g., battery acid) is very acidic, and a
pH of 13.0 (e.g., lye) is very alkaline.
The scale is logarithmic, so a pH of 4.0 is
10 times as acidic as pH 5.0 and 100 times as acidic as pH 6.0.

Source: The National Atmospheric Deposition Program and the
Canadian Network for Sampling Precipitation

focation. Dissolved aluminum also kills Mycorrhizol fungi which
is important to the health of most tree species.

Although all soils have some capacity to absorb acids, just
how much depends on the chemistry and thickness of the soil in
question. Some soils have a protective mechanism—calcium
carbonate, these are the limestone soils. This protective mechanism works on the chemical principle that acids and bases neutralize each other when mixed together. The limestone in the
soil reacts with the acid in the rainfall, neutralizing the acidity,
thereby lending protection to the lake water. By the time the rain
has filtered through the soil and reached the lake, the acidity has
been neutralized. Unfortunately, few if any of the Adirondack's
2,500 lakes possess this protective chemistry. At last count, the
Adirondack Mountain region in the Northeastern United States
has 200 lakes which are too acidic to support life and another
200 which show some damage. Canada, to the north, has 2000
dead lakes, and if the amount of air pollution from both Canada
and the United States are not sharply reduced, another 48,000
lakes will be threatened over the next 20 years. Presently, the
United States produces about five times more sulfur pollution
than Canada, sending about 2 million tons of sulfur dioxide per

154

year on the prevailing winds into Canada. And Canada sends about a half-million tons per year south.

Sweden

In Sweden, there are at least 3 to 4 thousand severely acidic lakes and another 14 to 15 thousand which are damaged. Out of its 90 thousand lakes, 18 to 20 thousand of them are affected by acid rain, numbers which indicate an advanced stage of ecological decline. Although both Sweden and Norway on the Scandinavian peninsula have adopted strict measures to reduce sulfur emissions from their own industries, their acid rain problem persists because the major portion of the acid rain affecting their environments originates in Great Britain, France, and Germany. Sweden is responsible for only about 22 percent of its sulfur burden, and Norway contributes, at most, 20 percent of its sulfur pollution. Europe's sulfur emissions remained at about 25 million tons yearly until 1950, however, since 1950 this level more than doubled reaching the level of 60 million tons yearly in 1973.

GENERAL ACCOUNTING OFFICE

Intense debate continues over whether it is time for the United States to take control actions against the environmental effects of acid deposition. Often referred to as "acid rain," this phenomenon occurs when oxides of sulfur and nitrogen emitted by coal-fired power plants, smelters, vehicles, and other sources, both man-made and natural, are transported in the atmosphere and return to earth as acid compounds. This has become a national and international issue because these substances can often be transported beyond the jurisdictions in which they are emitted, and possibly damage the environment across state and even national boundaries. . . .

Effects of Acid Deposition

Investigation of the effects of acid deposition has concentrated on lakes and fish, forests and agriculture, materials, and human health.

GAO found that, to date, the only thorough and convincing documentation of damage in North America concerns the acidification of some lakes and rivers and the consequent reduction or elimination of populations of certain fish species in three areas of the northeastern United States and southeastern Canada. Significant biological damage caused by acid deposition in the United States has thus far been identified only in the Adirondack Mountain lakes of New York, but there is evidence of acidification of lakes and streams in wider areas which may continue and result in further loss of fish populations. Observations in Norway, which experienced the complete loss or substantial reduction of fish populations in thousands of lakes because of acidification, suggest its potential seriousness. However, attempts to forecast the future extent of the problem in North America remain inconclusive because of the wide range in scientific predictions, which range from no further damage to greater and more wide-spread damage at presently expected levels of deposition. . . .

Reports of harm from metals dissolved in acidified drinking water are rare to date in the United States but this question needs wider examination since lead causes brain damage. If individual cisterns and/or wells are acidified, as has occurred in Scandinavia, protective treatment could be inconvenient and relatively expensive. . . .

Emission Reduction Techniques

The three emission reduction techniques most extensively considered are washing coal, "scrubbing" the flue gases from older power plants, and switching to lower sulfur fuels. GAO examined each of these alternatives for its effectiveness and economic as well as social consequences.

Coal washing, already used extensively, can remove about 20 to 35 percent of the sulfur from high-sulfur coals at low capital costs. . . .

Switching to lower sulfur coals or scrubbing are the two methods asserted capable of making larger reductions in eastern U.S. SO_2 emissions. Each method can, under favorable circumstances, reduce SO_2 emissions for costs that have been estimated as low as a few hundred dollars per ton of SO_2, with the lowest cost estimate for switching being less expensive than the lowest cost estimate for scrubbing. The cost and feasibility of both methods are strongly influenced, however, by a number of factors that are very specific to individual plants. . . .

Controlling Acid Deposition

Congressional decisions about acid deposition revolve around whether to begin control actions promptly or to wait until better scientific data are available. Closely following on that question, however, are two issues sufficiently important to influence decisions on this main issue—what kind of control actions would be best, and how, and by whom, should they be paid for?

GAO's principal observation is that, although science has largely determined the causes of acid deposition, there is uncertainty concerning the amount and the timing of the effects which can be anticipated from it. Thus, scientific information alone cannot determine whether it would be better to begin control actions now or wait until estimates of effects can be made more accurate. Instead, the issue must be approached by weighing the relative risks of alternative decisions: the risks of adverse economic impacts in some regions of the country, caused by immediate control actions whose benefits cannot be accurately evaluated, versus the risks of further, potentially avoidable harm to the environment and possibly to public health in other regions of the United States and/or Canada if actions are delayed. . . .

Techniques for Controlling the Effects of Acid Deposition

—Mitigation actions taken where deposition occurs, such as liming of lakes, can prevent damage is some cases. However, they have limited capabilities both because they cannot control all kinds of damage, and also because they could not be applied economically to large unmanaged areas such as forests.

—If deposition reduction is desired, to control the risk of damage stemming from acid deposition, the greatest reduction in risk would come from lessening the deposition of acidic sulfur compounds, which could be accomplished best by reducing SO_2 emissions.

—Because deposition at almost any location includes significant contributions from sources spread over a wide area, emission controls intended to produce substantial reductions of acid deposition, even at one location, would be needed over a wide area rather than at one source or a narrowly localized set of sources.

General Observations on the Acid Deposition Issue

—Because the Clean Air Act currently focuses on concentrations of pollutants near their sources, any air pollution control approach to deal with acid deposition in this century would necessitate additions to, or a basic reorientation of, the ambient air quality standard approach in the present act.

—The dispute persists over whether it would be advisable to establish emission controls promptly to reduce acid deposition or to wait further. However, at a minimum, having control plans ready could save time, and therefore spare resources, if/when a need for rapid action becomes evident.

—Further scientific work on acid deposition will be needed for a number of years, no matter what decisions are made on control actions in the short run.

CURB ACID RAIN WITH ACTION

Jacquelyn L. Tuxill

Jacquelyn L. Tuxill is the chairperson of the New Hampshire Citizens' Task Force on Acid Rain.

Points to Consider

1. How are public attitudes toward acid rain described?
2. What kind of pollution cutbacks are recommended?
3. Why must the acid rain control program be national in scope?
4. Why do they recommend immediate action to curb acid rain?

Reprinted from a position paper by the New Hampshire Citizens' Task Force on Acid Rain, February 10, 1984.

The people of this country will and do support an acid rain control program. Americans have a fundamental sense of fairness, they care about the integrity of their environment and natural resources, and they believe in the basic right to breathe clean air and drink clean water.

Former EPA Administrator William Ruckelshaus has called acid rain one of the most regionally divisive issues he has ever encountered. The Reagan Administration's position of more study and no action is the most divisive course we could follow on acid rain. It is also regressive. The *people* of this country have put divisiveness behind them and are actively seeking solutions to the acid rain problem, not denials and inaction.

The citizens' conference, ACID RAIN '84, held in Manchester, N.H. one month ago is illustrative of this fact. Nearly 700 people from all across the country and Canada, 150 of them representing the media, gathered together seeking remedies and *action* on acid rain. Coal miners from Kentucky sat down with sportsmen from New England. Representatives of the food products industry mingled with people concerned with public health effects of acid rain. Labor interests, educators, public officials, native Americans, legislators, scientists, and environmentalists from the midwest, the south, the northeast, the southwest and the far west—all committed to solving the problem *together*—listened to scientists talk of the urgent need to control acid rain, heard of growing evidence of acid rain damage in the midwest and Appalachia, discussed with experts the control options available, the costs, the impact on jobs. After two days there was not divisiveness and denials, but unanimous agreement and a shared commitment to doing what is fair. A citizens platform calling for a 50% reduction in sulfur emissions nationwide by 1990 was unanimously adopted, as was a U.S./Canada citizens agreement calling on both countries to commit themselves to the 50% SO_2 reduction goal.

Public Attitudes

I say to you as strongly as I can that the people of this country will and do support an acid rain control program. Americans have a fundamental sense of fairness, they care about the integ-

160

Moving Toward a Solution

The National Academy of Sciences recently endorsed the conclusions of work by the Environmental Defense Fund's staff physicist, Dr. Michael Oppenheimer, proving that acid rain can be reduced by curbing the emissions of coal-fired power plants. Nationwide, American industry puts about 26 million tons of sulfur dioxide into the air each year. Coal-fired power plants contribute nearly two-thirds of this total.

Hundreds of lakes in the Northeast United States are already wiped clean of fish by acid rain, along with hundreds of miles of streams. In many lakes restocking may be impossible because of permanent changes in the chemistry of the water and surrounding soil.

Tens of thousands of lakes and as many miles of streams are endangered, many already showing alarming decreases in their capacity to buffer the acid falling into them from the sky.

Environmental Defense Fund, 1984

rity of their environment and natural resources, and they believe in the basic right to breathe clean air and drink clean water.

Pollster Louis Harris has tapped into this deep vein of concern. In a speech before the Coalition of Northeastern Governors on December 4, 1983, Mr. Harris relayed the following results relating to acid rain:

• By an 86–12% margin, Americans want to keep the Clean Air Act intact or make it stricter.

• By 65–32%, Americans oppose relaxing standards on the basis of the cost of cleaning up air pollution that endangers human health.

• Today 63% of Americans are familiar with acid rain, up from 55% in 1982, 43% in 1981, and 30% in 1980.

• By 73–22%, Americans feel it is fair that the costs of cleaning up acid rain should be borne by all individuals and businesses who use fuels that contribute to the problem. Regional sentiments on this cost-sharing question were essentially all in the same range. . . .

161

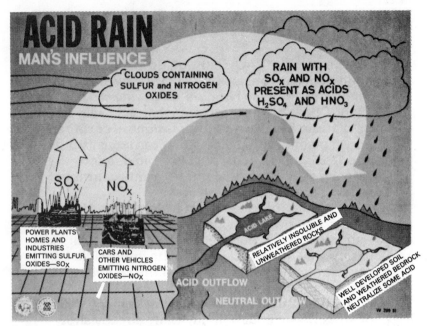

ACID RAIN
MAN'S INFLUENCE

CLOUDS CONTAINING
SULFUR and NITROGEN
OXIDES

RAIN WITH
SO_X AND NO_X
PRESENT AS ACIDS
H_2SO_4 AND HNO_3

SO_X NO_X

POWER PLANTS
HOMES AND
INDUSTRIES
EMITTING SULFUR
OXIDES—SO_X

CARS AND
OTHER VEHICLES
EMITTING NITROGEN
OXIDES—NO_X

ACID LAKE

RELATIVELY INSOLUBLE AND
UNWEATHERED ROCKS

WELL DEVELOPED SOIL
AND WEATHERED BEDROCK
NEUTRALIZE SOME ACID

ACID OUTFLOW

NEUTRAL OUTFLOW

Source: U.S. Geological Survey

We feel the following elements are essential to an effective acid rain control program:

1. **The SO$_2$ reduction must be 50% at a minimum.**

The National Academy of Sciences (NAS) in 1981 concluded that a 50% reduction in deposited hydrogen ions was necessary to "significantly reduce the rate of deterioration of sensitive freshwater ecosystems." Similarly, the January 1983 Canadian Summary of the U.S. Canada Working Group proposed that wet sulfate deposition be reduced to "less than 20 kg./hectar/yr. in order to protect all but the most sensitive aquatic ecosystems." In New Hampshire this translates into slightly more than a 50% reduction in sulfur loading.

Let us not lose sight of the fact, however, that even with a 50% SO$_2$ reduction, precipitation in the northeast will still be acid. The 50% figure is not a magic threshold between damage and no damage. It should reduce acidity levels, though, to the point (greater than pH 4.5) where aquatic ecosystems can support fish life.

If a control program of less than 50% SO$_2$ reduction is passed, the damage process may be slowed some, but the end result will be the same. We in New Hampshire, who regularly experience some of the most acidic precipitation in the country, feel

162

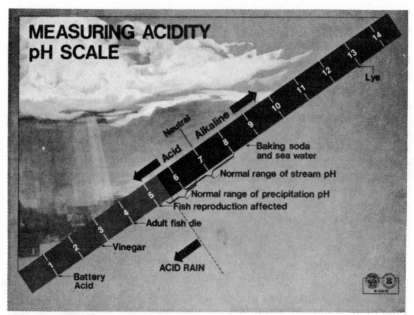

MEASURING ACIDITY
pH SCALE

Neutral
Acid
Alkaline

14
13
12
11
10
9
8
7
6
5
4
3
2
1

Lye

Baking soda
and sea water

Normal range of stream pH

Normal range of precipitation pH

Fish reproduction affected

Adult fish die

Vinegar

ACID RAIN

Battery
Acid

Source: U.S. Geological Survey

most strongly that 50% SO$_2$ reduction is the absolute, non-negotiable minimum.

2. The acid rain control program must be national in scope.

Two years ago when the Senate was wrestling with this issue the main areas of recognized damage were the upper midwest and the northeast. As more research has been done, more people brought into the decision-making process, and the solutions thought through in greater depth, it has become increasingly clear that acid rain is a problem national and international in scope. Because of the regional inequities, international implications, the potential for economic dislocations inherent in the fuel-switching debate, and the growing evidence of acid rain damage in areas other than the northeast, the solution must be national in scope and it must come now.

Many of these sub-issues are admittedly thorny; they are, however, definitely solvable. The Acid Rain Task Force of the National Governors' Association (NGA) has tackled the issue and recommended a national program involving significant SO$_2$ reductions. The NGA's executive committee is currently circulating a proposal for 10 million ton SO$_2$ reduction. Endorsements for acid rain control have also come from the Coalition of Northeastern Governors, the New England Governors Conference, the

Western Governors Policy Office, State and Territorial Air Pollution Program Administrators, and the National League of Cities.

3. **Any cost-sharing program incorporated into acid rain legislation must be tightly coupled with the 50% SO$_2$ reduction target.**

While the Senate does not currently have a bill incorporating cost sharing, the concept is worth discussing. We in New Hampshire do not want to pay into a program that will not be effective in the long run in controlling the acid rain damage we're experiencing. We would essentially be paying others to despoil our resources. In the interest of arriving at a solution to the problem, however, we are willing to consider a cost-sharing program that requires a 50% SO$_2$ reduction nationwide and takes into account past SO$_2$ cleanup efforts, and current contributions to total sulfur emissions. The cost-sharing proposal in HR4404, introduced into the House last November by Rep. Norman D'Amours is worth consideration in this respect.

The state is set for action. The scientific evidence to justify immediate control of acid rain is there. The support of the citizens of this country for a fair and equitable acid rain control program is there and the commitment will continue to grow, I can assure you. But the leadership from the Reagan Administration is not there.

MORE STUDY NEEDED ON ACID RAIN ISSUE

William D. Ruckelshaus

William D. Ruckelshaus was formerly the administrator of the Environmental Protection Agency in the first Reagan Administration.

Points to Consider

1. What are the four major gaps in our knowledge about acid rain?
2. Why is more study needed before major actions are taken?
3. What kind of research program is taking place?
4. When will enough knowledge exist to begin a major national effort to control acid rain?

"William Ruckelshaus Speaks Out on Acid Rain," *Conoco,* Volume 15, 1984.

Before large sums of money are committed to controls, we should have as clear a picture as possible of what we are getting for this investment.

There is no question that there is an acid rain problem in this country. But there are questions about the scope of that problem, the pace at which damage may be increasing, the precise way damage occurs, and the best ways to correct it. There are gaps in our knowledge in at least four areas that are relevant to forming a sound acid rain policy.

The first, and perhaps most basic gap, involves the scope of the problem. We really do not know the extent of the damage presently caused by acid deposition. A limited number of lakes have been surveyed and some of those have been found acidic, but even here, no comprehensive inventory exists.

Second, we don't know at what pace this observed damage has occurred. Knowing whether we face a decades-long problem or a more rapidly deteriorating situation is important for the development of sound policies. We do know that a major contributor to the acidity of rainfall, sulfur oxides, has declined by more than 15 percent since the passage of the Clean Air Act in 1970. Sulfur oxides are projected to remain at roughly current levels until the turn of the century, if we impose no additional controls.

Third, we are uncertain whether or to what extent present levels of damage may be getting worse as a result of current levels of emissions. While our data are sparse, they tend to show that the acidity of rainfall in the eastern United States has stayed essentially level for the past 20 years. If it is the rate of deposition that determines the extent of damage in particular watersheds, those watersheds may have reached a steady state in which no more damage will occur unless emissions increase, since the rate of deposition has held steady or decreased over the last few years in the Northeast. If it is the accumulated deposition that matters for a watershed, then damage may still be increasing, even though the rate of deposition remains the same.

At my request, a panel of the National Academy of Sciences has reviewed this question of the acidification process for surface waters. The preliminary conclusion is that continued deposition at the current rate is unlikely to have any additional impact in watersheds where the soil has a high ability to neutralize acid deposition, or in watersheds whose rate of acidification is at equilibrium with the rate of acid deposition. It may matter for those watersheds where the soil's neutralizing capacity is moderate and not readily renewed and thus might be exhausted by

166

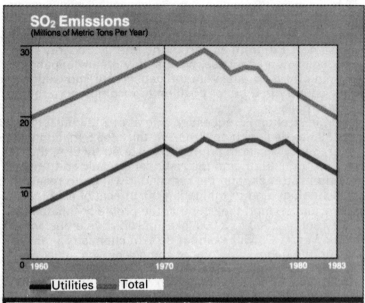

SO₂ Emissions
(Millions of Metric Tons Per Year)

Utilities — Total

The rapid increase in SO₂ emissions during the 1960s and early 1970s has been reversed. Both total SO₂ emissions and those from utilities were over 20 percent less last year than during the peak year of 1973.

Source: EPA

repeated stress from acid rain. We have very little idea how many watersheds fall in this category.

Fourth, and finally, there are gaps in our knowledge of where the acid deposition in any given area comes from. This is the source-receptor problem. Our current atmospheric models cannot adequately predict the impact of particular sources on particular receptor areas. Further, though much acid deposition comes down in a dry form, the techniques to measure this dry deposition are still under development.

These uncertainties and gaps in our knowledge are matters of far more than academic interest. They are directly relevant to responsible public policy choices.

Any acid deposition control program will represent a major environmental, economic, and social investment for this country. Knowledge of how and where to target that program, how big to make it, and what to expect from it, will be crucial in assuring that the investment is wisely made. Any control program will be costly. Before large sums of money are committed to controls, we should have as clear a picture as possible of what we are getting for this investment.

Expanding Research

In December, 1980, Congress set up the National Acid Precipitation Assessment Program as part of the Energy Security Act. With that legislation as a base, we are now greatly expanding our research efforts to deal with the gaps in our knowledge and to put our country in a better position to recommend targeted and efficient policies.

The Administration is requesting a total of $127 million in fiscal year 1985 to deal with acid rain. Of this, $55.5 million will go to an interagency research effort administered by EPA, the Department of Agriculture, and the National Oceanic and Atmospheric Administration, and the remainder will go to research on control technology and to mitigating the effects of acid rain.

To better understand the scope of the problem, this fall we will survey some 2,000 to 3,000 lakes in sensitive areas. In succeeding years, we will examine stream chemistry in these regions and the general biology of some representative watersheds.

Our national trends network, which monitors wet deposition, has been put in place and will be expanded in the next few years. We are also developing a monitoring network for dry deposition. These efforts should give us a much better idea both of the nature and extent of total deposition, and will eventually reveal long-term deposition trends.

As I mentioned earlier, the National Academy of Sciences is assisting us in reviewing the scientific evidence regarding the mechanisms of acidification of surface waters. I have asked the panel to recommend additional research approaches to help us clarify our understanding of these mechanisms.

The Federal Interagency National Acid Precipitation Assessment Program will concentrate on three areas to improve our understanding of source/receptor relationships. First, we will conduct field tracer studies to obtain a better understanding of the complex meteorology of long-distance pollutant transport. Second, we will seek to better characterize the complicated sequence of chemical reactions which makes up the transformation process. (That is, the process by which sulfates, nitrates, oxidants, and other pollutants are formed in the atmosphere.) Finally, we will incorporate this improved understanding of meteorology and chemistry into more sophisticated atmospheric models so that we can analyze and predict the effectiveness of alternative control strategies.

Forests and Health

Now, let me turn to two areas that have only recently assumed more importance in the acid rain debate and that have received
168

considerable attention. These are damage to forests and damage to health.

Based on limited data, it appears that over a wide area of the eastern United States there has been a pronounced decline in tree diameter growth of several species of trees over the past two decades. . . . In Europe, different and more extensive types of tree damage have been observed, involving at least ten species.

We do not know the true extent or meaning of this damage, the speed at which it is taking place, or what factor or combination of factors is causing it. We do not know if the causes are the same in Europe as in this country. . . .

This situation illustrates well why waiting for further research to be completed before initiating a control program is a rational decision. If, as many believe, sulfate deposition is not a major contributor to forest problems, but oxidants or nitrates are, a significant reduction in sulfur dioxide (SO_2) emissions could inadvertently result in elevated levels of oxidants or nitrates. Our current understanding of atmospheric chemistry indicates that, if we were to reduce SO_2, it might result in increased levels of oxidants. Additionally, excess oxidant could then combine with oxides of nitrogen to produce more nitrates. Thus, in either case controlling the wrong pollutant could conceivably make matters worse. . . .

As to health, some witnesses have testified before Congress that acid rain causes health damage. We agree with the recent National Institute of Environmental Health Sciences report on this subject which, though cautious, did not find any basis for immediate alarm. Nevertheless, as the Institute suggests, further assessment is warranted to expand our understanding of such potential effects as the leaching of heavy metals into drinking water by acid rain and the impacts of breathing sulfates and acid fog.

169

We Can Solve It

I have been repeatedly asked when we will know enough to make a decision regarding controls. The answer is that I do not know, because I cannot predict ahead of time what answers will come out of our research program—or when.

The Interagency Task Force plans to produce formal assessments of the information gained from the acid deposition research program in 1985, 1987, and 1989. These will be important milestones in integrating our understanding of acid rain's causes and effects. As we continue to gain knowledge of the deposition problem, our ability to predict the results of various control efforts will increase, and we will reach the point where the Administration can responsibly make a decision regarding the need for additional controls. I cannot tell exactly when that point will come, for I cannot predict what the answers from the research program will be or when they will be forthcoming. What I can say is that I take it as an affirmative duty on my part as administrator of EPA to ensure that we make this active reassessment an ongoing process and that I communicate our newly found knowledge to the key decision-makers in the Administration, including the President, as soon as appropriate. I believe that, if we all approach the acid rain problem with good will and a recognition of the legitimate concerns of people in every section of this country and Canada, we can solve it.

INTERPRETING EDITORIAL CARTOONS

This activity may be used as an individualized study guide for students in libraries and resource centers or as a discussion catalyst in small group and classroom discussions.

Although cartoons are usually humorous, the main intent of most political cartoonists is not to entertain. Cartoons express serious social comment about important issues. Using graphic and visual arts, the cartoonist expresses opinions and attitudes. By employing an entertaining and often light-hearted visual format, cartoonists may have as much or more impact on national and world issues as editorial and syndicated columnists.

Points to consider

1. Examine the two cartoons in this activity.
2. How would you describe the message of each cartoon? Try to describe each message in one to three sentences.

Heller, West Bend News

3. Do you agree with the message expressed in either cartoon? Why or why not?
4. Do either of the cartoons support the author's point of view in any of the readings in this publication? If the answer is yes, be specific about which reading or readings and why.
5. Which reading in chapter four would be in basic agreement with the message of the Carol and Simpson cartoon?

BIBLIOGRAPHY

Government Reports And Studies

"Assessment of Hazardous Waste Mismanagement: Damage Case Histories," December, 1982. **Synopsis:** This Study was designed to assess, using the available data bases, the types of damages that could be expected to occur from mismanaging hazardous wastes. **Source:** Office of Solid Waste/U.S. Environmental Protection Agency/401 M Street S.W./ Washington, DC 20460/(202)382-4667.

California: "[DHS] Reports Containing Groundwater Contamination Information." **Synopsis:** In February 1983, the Department of Health Services provided information on groundwater contamination by organic chemicals and pesticides. **Source:** Environmental Health Division/California Department of Health Services/2151 Berkeley Way/Berkeley, California 94704/(415)540-2172.

"Community Water Supply Survey" [CWSS], 1978. **Synopsis:** This report provides information on community water supplies from around the country. **Source:** Office of Drinking Water/U.S. Environmental Protection Agency/401 M Street, SW/Washington, D.C. 20460.

"Compendium of Cases of Ground Water Contamination," 26 August, 1982. [140 pages] **Synopsis:** This report was prepared for the use of the Groundwater Policy Group of EPA to assist them in formulating an agency groundwater strategy. **Source:** Economic and Policy Analysis Branch/Office of Drinking Water (WH-550)/U.S. Environmental Protection Agency/401 M Street, SW/Washington, D.C. 20460.

"Contamination of Ground Water by Toxic Organic Chemicals," January, 1981. [84 pages] **Synopsis:** This report is divided into three chapters: 1. Groundwater Characteristics and Contamination, 2. Contamination of Drinking Water Wells, and 3. The Health Risks of Toxic Organic Chemicals. **Source:** Council On Environmental Quality/722 Jackson Place, N.W./Washington, D.C. 20006.

"Damages and Threats Caused by Hazardous Material Sites," February, 1980. [175 pages] **Synopsis:** This report lists 244 hazardous material incidents from 41 States. **Source:** Office of Water and Waste Management/U.S. Environmental Protection Agency/401 M Street, SW/Washington, D.C. 20460.

"Drinking Water Supplies Contaminated by Organic Chemicals in New England," 7 March 1983. [21 pages] **Synopsis:** This listing is periodically updated and provides information exclusively on drinking water wells known to be contaminated by organic chemicals in the New England States. **Source:** Office of Water Supply, Region I/U.S. Environmental Protection Agency/Boston, Massachusetts 02203/(617) 223-4600.

"Florida Ground Water Contamination Sites," 26 August 1982. [31 pages] **Synopsis:** This inventory contains 43 cases found in 19 counties. Paragraph-long descriptions accompany each case. **Source:** Groundwater Section/Florida Department of Environmental Regulation/Twin

Towers Office Building/2600 Blair Stone Road/Tallahassee, Florida 32301-8241/(904) 488-3601.

"Identification of Organic Compounds in Effluents from Industrial Sources," April, 1975 [Final Report: EPA-560/3-75-002] [248 pages]. **Synopsis:** This report contains useful information on organic chemicals, their reactions in water and air, chemical classifications, and biodegradability. **Source:** Office of Toxic Substances/U.S. Environmental Protection Agency/401 M Street, SW/Washington, D.C. 20460.

"National Ground Water Supply Survey" [GWSS], June, 1982. **Synopsis:** The survey conducted a sampling and analysis program on finished water from 945 public water supply systems which depend on groundwater sources. **Source:** Technical Support Division/Office of Drinking Water/U.S.Environmental Protection Agency/Cincinnati, Ohio 45268.

"National Water Data Exchange [NAWDEX]". **Synopsis:** A primary objective of the NAWDEX is to improve access to water data and to help users identify, locate and acquire needed data. **Source:** U.S.Geological Survey/Box 25425, Federal Center/Lakewood, Colorado 80225/(303) 234-5888.

New York: "Organic Chemicals and Drinking Water," April, 1981 [140 pages]. **Synopsis:** This comprehensive report, revised from 1980, includes information on groundwater and selected cases of surface water contamination. **Source:** New York State Department of Health/Empire State Plaza, Tower Building/Albany, New York 12237/(518) 474-5577.

"Occurrence of Organic Chemicals in Drinking Water, Air and Food," January, 1983. [completed in 1982] **Synopsis:** This report summarizes information on trichloroethylene [TCE] in drinking water; it is one in a series of studies on the presence of 13 volatile organic chemicals [VOC] in drinking water. **Source:** Office of Drinking Water/U.S.Environmental Protection Agency/401 M Street, SW/Washington, D.C. 20460/(202) 382-3022.

Pennsylvania: "Water Quality Inventory, 1982, "April, 1982 [318 pages]. **Synopsis:** This document reports on all facets of the State water quality management program and was prepared pursuant to requirements of Section 305(b) of the Clean Water Act. **Source:** Bureau of Water Quality Management/Department of Environmental Resources/Commonwealth of Pennsylvania/P.O. Box 2063/Harrisburg, Pennsylvania 17120.

Groundwater Contamination: Selected References

Burmaster, David E. The new pollution; groundwater contamination. **Environment,** v. 24, Mar. 1982: 6–13, 33–36.

Canter, Larry W. Groundwater quality management. **American Water Works Association Journal,** v. 74, Oct. 1982: 521–527.

Ferrett, Robert L., and Robert M. Ward. Agricultural land use planning and groundwater quality. **Growth and Change,** v. 14, Jan. 1983: 32–39.

Hoberg, Allen C. The nature and extent of groundwater management and groundwater problems: a survey. **Agricultural Law Journal,** v. 4, fall 1982: 404–442.

Horne, Amy. Groundwater policy: a patchwork of protection. **Environment,** v. 24, Apr. 1982: 6–11, 35.

Jackson, R. E. The contamination and protection of aquifers. **Nature and Resources,** v. 18, July–Sept. 1982: 2–6.

Josephson, Julian. Protecting public groundwater supplies. **Environmental Science and Technology,** v. 16, Sept. 1982: 502A-505A.

_____ Restoration of aquifers. **Environmental Science & Technology,** v. 17, Aug. 1983: 347A–350A.

Marinelli, Janet. It came from beneath Long Island. **Environmental Action,** v. 14, May 1983: 8–12.

Mosher, Lawrence. Polluted groundwater clearly a problem, but few agree on extent or solution. **National Journal,** v. 16, Feb. 2, 1984: 223–225.

Pye, Veronica I., and Ruth Patrick. Ground water contamination in the United States. **Science,** v. 221, Aug. 19, 1983: 773–781.

Pye, Veronica I., Ruth Patrick and John Quarles. Groundwater contamination in the United States. Philadelphia, University of Pennsylvania Press, 1983. 315 p.

Reitman, Frieda. Costs and benefits in aquifer protection. **New England Journal of Business & Economics,** v. 9, fall 1982: 41–49.

Tangley, Laura. Groundwater contamination: local problems become national issue. **BioScience,** v. 34, Mar. 1984: 142–146, 148.

Taylor, Ronald A. Warning: your drinking water may be dangerous. **U.S. News & World Report,** v. 96, Jan. 16, 1984: 51, 54.

U.S. Environmental Protection Agency. Ground-Water Task Force. A groundwater protection strategy for the Environmental Protection Agency; draft. Washington, The Agency, 1984. 1 v. (various pagings).

Williamson, Dayle E., and others. Point . . . and counterpoint: [groundwater contamination] **Environmental Forum,** v. 2, Jan. 1984: 26–32.

Workshop on groundwater resources and contamination in the United States (summary and papers). Washington, National Science Foundation [available from NTIS] 1983. 237 p.

Yeany, Philip R. Permit fees for New Jersey's surface and groundwater dischargers. **Environmental Forum,** v. 2, Jan. 1984: 21–25.

Clean Water Act: Selected References

Clean water program gets mixed review. **Engineering News-Record,** v. 208, Jan. 14, 1982: 10–11.

Cullen, Michael J., Clyde H. Burnett, and James A. Chamblee. Total domestic wastewater costs pegged at $22 billion a year. **Journal [of the] Water Pollution Control Federation,** v. 53, May 1981: 522–529.

Marth, Del. A whirlpool of waste. _Nation's Business,_ v. 69, Aug. 1981: 51–53.

Mosher, Lawrence. Clean water requirements will remain even if the Federal spigot is closed. **National Journal,** v. 13, May 16, 1981: 874–878.

Other Water Quality Issues

Blumm, Michael C. The Clean Water Act's section 404 permit program enters its adolescence: an institutional and programmatic perspective. **Ecology Law Quarterly,** v. 8, no. 3, 1980: 409–472.

Chapman, William. Environmental groups assail EPA on its clean-water proposals. **Washington Post,** Apr. 9, 1982: A9.

Clean water: why industry won't get what it wants. **Chemical Week,** v. 130, Mar. 31, 1982: 28–31.

Kerner, Michael P. Highlights of the Clean Water Act of 1977. **Environmental Law,** v. 8, spring 1978: 869–885.

Koch, Kathy. Congress to review clean water legislation. **Congressional Quarterly Weekly Report,** v. 40, Jan. 23, 1982: 124.

Mosher, Lawrence. Reagan's environmental Federalism—are the states up to the challenge? **National Journal,** v. 14, Jan. 30, 1982: 184–188.

Myhrum, Christopher B. Federal protection of wetlands through legal process. **Boston College Environmental Affairs Law Review,** v. 7, no. 4, 1979: 567–627.

The Pitfalls of 'defederalization.' **Chemical Week,** v. 128, June 17, 1981: 54–58.

Rastatter, Clem L. Congress braces for new fight on water quality. Conservation Foundation Letter, Sept. 1981: 1–8.

Shabecoff, Philip. Administration seeks eased rules for industries in Clean Water Act. **New York Times,** Apr. 9, 1982: A1, A28.

Additional Reference Sources

Abrams, E. F., D. Derkics, C. V. Fong, D. K. Guinan, and K. M. Slimak, "Identification of Organic Compounds in Effluents From Industrial Sources," NTIS PB-241641, 1975.

Burmaster, D. E., and R. H. Harris, "Groundwater Contamination: An Emerging Threat," **Technology Review** 85(5):50–62, 1982.

Coniglio, W., "Criteria and Standards Division Briefing on Occurrence/Exposure to Volatile Organic Chemicals," Office of Drinking Water, U.S. Environmental Protection Agency, 1982.

Considine, D. M., and G. D. Considine (eds.), **Van Nostrand's Scientific Encyclopedia,** 6th ed. (New York: Van Nostrand Reinhold, 1983).

Council on Environmental Quality, "Contamination of Ground Water by Toxic Organic Chemicals," Washington, DC, 1981.

Environ Corp., "Approaches to the Assessment of Health Impacts of Groundwater Contaminants," draft report prepared for the Office of Technology Assessment, August 1983.

Fryberger, J., "Rehabilitation of a Brine-Polluted Aquifer," Office of Research and Monitoring, U.S. Environmental Protection Agency, 1972.

Harris, R. H., "The Health Risks of Toxic Organic Chemicals Found in Groundwater," Princeton University/Center for Energy and Environmental Studies, Report No. 153, 1983.

Hawley, G. G., **The Condensed Chemical Dictionary,** 9th ed. (New York: Van Nostrand Reinhold, 1977).

Miller, D. W. (ed.), **Waste Disposal Effects on Ground Water** (Berkeley, CA: Premier Press, 1980).

National Academy of Sciences, **Drinking Water and Health** (Washington, DC: National Academy Press, 1977).

National Academy of Sciences, **Risk Assessment in the Federal Government: Managing the Process** (Washington, DC: National Academy Press, 1983).

O'Brien, R. P., and J. L. Fisher, "There is an Answer to Groundwater Contamination," **WATER/Engineering and Management,** p. 30ff, May 1983.

Odum, E. P., **Fundamentals of Ecology,** 3d ed. (Philadelphia: W. B. Saunders Co., 1971).

Patterson, W. L., and R. F. Banker, "Effects of Highly Mineralized Water on Household Plumbing and Appliances," paper presented at the Annual Conference, American Water Works Association, June 1968.

Prichard, H. M., and T. F. Gesell, "Radon-222 in Municipal Water Supplies in the Central United States," **Health Physics** 45:991–993, 1983.

Radike, M. J., K. L. Stemmer, P. G. Grown, E. Larson, and E. Brigham, "Effect of Ethanol and Vinyl Chloride on the Induction of Liver Tumors: Preliminary Report," **Environmental Health Perspectives** 21:153–155, 1977.

Raucher, R. L., "A Conceptual Framework for Measuring Benefits of Groundwater Protection," **Water Resources Research** 19:320–326, 1983.

Reitman, F., "Costs and Benefits in Aquifer Protection," **New England Journal, Business and Economics** 19(1), 1982.

Sheridan, D., "Desertification of the United States" (Washington, DC: Council on Environmental Quality, 1981).

U.S. Environmental Protection Agency, "Hazardous Waste Disposal Damage Reports," No. 3, 1976.

APPENDIX . . . Sources of Groundwater Contamination*

1. Subsurface Percolation—Septic Tanks and Cesspools: There were an estimated 19.5 million *domestic* on-site disposal systems in the United States in the mid-1970s, of which 16.6 million were septic tanks and cesspools, presumably the remaining 2.9 million systems were privies or chemical toilets. Of all the sources known to contribute to groundwater contamination, septic tank systems and cesspools directly discharge the largest volume of wastewater into the subsurface.
2. Injection Wells: Several types of injection wells are used to inject or discharge wastes into or perform other functions in the subsurface: ● hazardous waste wells; ● non-hazardous waste wells (e.g., brine injection wells, and agricultural, urban runoff, and sewage disposal wells); and ● non-waste wells (e.g., wells for enhanced oil recovery, artificial recharge, in-situ recovery, and solution mining). **3. Land Application:** Land application of treated wastewater and wastewater byproducts (i.e., sewage sludge) is often used in place of more costly disposal processes. Its primary goals are the biodegradation, immobilization, and/or stabilization of various chemicals, and the beneficial use of nutrients contained in the wastewater or sludge. **4. Landfills:** The solid wastes deposited in landfills are generally classified as hazardous or non-hazardous. Hazardous solid wastes are specifically defined under RCRA regulations. **5. Open Dumps:** A dump is a land disposal site where solid wastes are deposited indiscriminately, with little or no regard for the design, operation, maintenance, or esthetics of the site. In an "open" dump, the wastes are almost always left uncovered. Most often the open dump is not authorized and there is no supervision of dumping. **6. Residential (Local) Disposal:** A variety of hazardous and toxic substances are commonly found in household wastes. These wastes often are disposed of in specific facilities designed for waste disposal or discharge (e.g., municipal landfills). However, they also are disposed of indiscriminately, without supervision, in gutters, sewers, storm drains, and backyard burning pits. **7. Surface Impoundments:** Surface impoundments are used by both industries and municipalities for the retention, treatment, and/or disposal of both hazardous and non-hazardous liquid wastes. They can be either natural depressions or artificial holding areas (e.g., excavations or dikes). **8. and 9. Waste Tailings and Waste Piles:** Mining operations generate two basic types of solid wastes—spoil piles and tailings. Spoil piles are generally disturbed soil and overburden from surface mining or waste rock from underground mining operations (Miller, 1980). Tailings are the solid wastes from the on-site operations of cleaning and extracting ores. **10. Materials Stockpiles:** Very little information has been obtained regarding either the numbers or the amounts of materials in stockpiles in the United States. Approximately 3.4 billion tons of various materials (e.g., coal, sand and gravel, crushed stone, copper ore, iron ore, uranium ore, potash, tita-

*Reprinted from "Protecting The Nation's Groundwater From Contamination," Vol. II., Office of Technology Assessment, October, 1984.

nium, phosphate rock, and gypsum) were produced in 1979 (Koch, et al., 1982). **11. Graveyards:** Decomposing bodies in graveyards produce fluids that can leak to underlying groundwater, especially if nonleak-proof caskets are used. **12. Animal Burial:** There are no data to assess the potential contribution of this source to groundwater contamination. **13. Aboveground Storage Tanks:** Aboveground storage tanks are used in industrial, commercial, and agricultural operations and at individual residences for a large variety of chemicals. No systematic information is available regarding numbers, sizes, and locations of these tanks or of the chemicals stored in them. **14. Underground Storage Tanks:** Underground storage tanks are used by industries, commercial establishments, and individual residences for storage and treatment of products or raw materials. **15. Containers:** Containers are storage barrels and drums for various waste and non-waste products. They can be moved around with relative ease, and although they may be buried, they are not specifically designed to be. **16. Open Burning and Detonation Sites:** Very little information is available on this source. Although there are probably many cases of waste materials burned in backyards or at landfills, these cases are classified here under the open dump, residential disposal, or landfill sources. **17. Radioactive Disposal Sites:** Radioactive materials arise from the nuclear fuel cycle, commercial and industrial products and wastes, and natural sources. They may have long half-lives, and they can migrate with no visible evidence. **18. Pipelines:** Pipelines are used to transport, collect, and/or distribute both wastes and non-waste products. **19. Material Transport and Transfer Operations:** Material transport and transfer operations refer to the movement of substances by vehicle (e.g., truck and railroad) along transportation corridors. Estimates of the number of spills vary. It is believed that yearly approximately 14 million tons of hazardous materials are spilled during material transport and transfer operations. This estimate is only a first approximation. **20. Irrigation Practices:** Water used for irrigation tends to percolate into the subsurface and move toward discharge points. As it does, it carries with it substances applied to and associated with the soil (e.g., fertilizers, pesticides, and sediment). **21. Pesticide Applications:** Pesticides are chemicals used for control of insects, fungi, and other undesirable organisms and weeds. **22. Fertilizer Applications:** Farmers used 54.0 million tons of commercial fertilizers in 1980–81, 48.7 million tons in 1981–82, and 42.3 million tons in 1982–83. **23. Animal Feeding Operations:** In the last two decades the number of animal feedlots with more than 1,000 animals has increased rapidly. **24. De-Icing Salts Applications:** Highway de-icing salts are applied to snow and ice-covered roads to improve driving conditions. Use of highway de-icing salts is confined primarily to the snowbelt. **25. Urban Runoff:** Urbanization necessarily expands the areas that are impervious to rainfall and thus increases the amount and rate of surface runoff. **26. Percolation of Atmospheric Pollutants:** Many potential contaminants of groundwater are carried in the atmosphere and eventually reach the land surface through either dry deposition between storms or transport in water and snow during storms. **27. Mining and Mine Drainage:** Deep underground mines, especially for coal, are located primarily in the Appalachian region; and surface mines are primarily in the West and Midwest. **28. Production Wells:** All production wells share a similar potential to contaminate groundwater. It is related to installation and operation methods (e.g., for oil wells, the use of treatment chemicals,

drilling fluids, and other chemicals), incorrectly plugged or abandoned wells, cross-contamination, and overdraft. **29. Other Wells:** Other wells include those used in various monitoring and exploration activities. No systematic information is available regarding numbers and locations of these wells. **30. Construction Excavation:** Excavation at construction sites can produce potential groundwater contaminants in a variety of ways. Heavy construction equipment used for rough grading can spill diesel fuel, oil and lubricants. **31. Groundwater—Surface Water Interactions:** When groundwater aquifers are hydrologically connected with surface water, the aquifer can be partially recharged by infiltration of the surface water. If the surface water is contaminated, or if it reacts chemically with the subsurface materials as it infiltrates downward, degradation of groundwater quality can follow. **32. Natural Leaching:** Natural leaching occurs on a local scale in aquifers, or in portions of aquifers, whose geologic materials can be dissolved into solution.
33. Salt-Water Intrusion/Brackish Water Upcoming: Approximately 21 billion gallons of groundwater per day—26 percent of all groundwater withdrawn (USDA, 1981a)—are withdrawn in excess of recharge capabilities. Overdrafting can disrupt the natural hydrologic processes associated with groundwater; and subsequent impacts on aquifers and groundwater quality include: salt-water intrusion in coastal areas, brine-water intrusion (or brackish water upcoming) in inland areas, and intensified natural leaching.